ATLANTIS STUDIES IN PROBABILITY AND STATISTICS

VOLUME 3

SERIES EDITORS: CHRIS P. TSOKOS

Atlantis Studies in Probability and Statistics

Series Editors:

Chris P. Tsokos

University of South Florida Tampa, Tampa, USA

(ISSN: 1879-6893)

Aims and scope of the series

The Series 'Atlantis Studies in Probability and Statistics' publishes studies of high-quality throughout the areas of probability and statistics that have the potential to make a significant impact on the advancement in these fields. Emphasis is given to broad interdisciplinary areas at the following three levels:

(I) Advanced undergraduate textbooks, i.e., aimed at the 3rd and 4th years of undergraduate study, in probability, statistics, biostatistics, business statistics, engineering statistics, operations research, etc.;

(II) Graduate level books, and research monographs in the above areas, plus Bayesian, nonparametric, survival analysis, reliability analysis, etc.;

(III) Full Conference Proceedings, as well as Selected topics from Conference Proceedings, covering frontier areas of the field, together with invited monographs in special areas.

All proposals submitted in this series will be reviewed by the Editor-in-Chief, in consultation with Editorial Board members and other expert reviewers

For more information on this series and our other book series, please visit our website at:

www.atlantis-press.com/publications/books

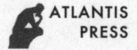

ATLANTIS PRESS

AMSTERDAM – PARIS – BEIJING

© **ATLANTIS PRESS**

An Introduction to Order Statistics

Mohammad Ahsanullah

Rider University,
Department of Management Sciences,
2083 Lawrenceville Road,
Lawrenceville, NJ 08648, USA

Valery B. Nevzorov

St. Petersburg State University,
Department of Mathematics and Mechanics,
198904 St. Petersburg, Russia

Mohammad Shakil

Miami Dade College (Hialeah Campus),
Department of Mathematics, 1800 West 49th Street,
Miami, FL 33012, USA

ATLANTIS
PRESS

AMSTERDAM – PARIS – BEIJING

Atlantis Press

8, square des Bouleaux
75019 Paris, France

For information on all Atlantis Press publications, visit our website at: *www.atlantis-press.com*

All books in this series are published in collaboration with Springer.

Atlantis Studies in Probability and Statistics

Volume 1: Bayesian Theory and Methods with Applications - Vladimir P. Savchuk, C.P. Tsokos
Volume 2: Stochastic Differential Games. Theory and Applications - K.M. Ramachandran, C.P. Tsokos

ISBNs
Print: 978-94-6239-048-5
E-Book: 978-94-91216-83-1
ISSN: 1879-6893

To my wife, Masuda

M. Ahsanullah

To my wife, Ludmilla

Valery B. Nevzorov

To my parents & my wife, Nausheen

M. Shakil

Preface

Dear Reader, imagine that you are a participant of Olympic Games, say, you are one of n competitors in high-jumping. Before the start the future results of participants can be regarded as some independent random variables $X_1, X_2, \ldots, X_n$. The competition will range all attempts of sportsmen and their final results can be considered as the observed values of the so-called order statistics $X_{1,n} \leqslant X_{2,n} \leqslant \cdots \leqslant X_{n,n}$. Hence to predict the result of the winner you must know the distribution of the extreme order statistic $X_{n,n}$. The future destinations of the silver and bronze prizewinners are determined as $X_{n-1,n}$ and $X_{n-2,n}$ correspondingly. If you are a sprinter then the future results of the gold, silver and bronze medaled sportsmen are associated with minimal order statistics $X_{1,n}$, $X_{2,n}$, and $X_{3,n}$. These are the simplest examples of the "sport" applications of order statistics. Other examples of the applicability of order statistics (and especially of extreme order statistics) can be suggested by meteorologists, hydrologists, business analysts. The knowledge of the theory of order statistics is useful for specialists in the actuarial science and the reliability theory.

Some attempts to present a systematic exposition of the theory of order statistics and extremes began essentially from the publication of the David's (1970) (the second issue of it appeared in 1981). We can mention also the following books, where the theory of order statistics and their different applications were presented: Galambos (1978, 1987), Arnold, Balakrishnan and Nagaraja (1992, 1998), Kamps (1995), Nevzorov (2000), Ahsanullah and Nevzorov (2001, 2005), David and Nagaraja (2003), Ahsanullah and Kirmani (2008). Almost all of these books are rather theoretical. We suggest here (see also Ahsanullah and Nevzorov (2005)) another way to study this theory. Together with the corresponding theoretical results, which are presented as 21 chapters, we suggest our readers to solve a lot of exercises. From one side it allows to understand better the main ideas and results of the theory. From other side the reader can determine his/her level of permeation to this material. Solutions of these exercises are given in the end of the corresponding chapters.

The aim of the book is to present various properties of the order statistics and inference based on them. The book is written on a lower technical level and requires elementary knowledge of algebra and statistics. The first chapter describes some basic definitions and properties of order statistics. Chapters 2 to 4 present sample quantiles, representation of order statistics as functions of independent and identically distributed random variables, conditional distributions and order statistics of discrete distributions. Chapters 5 to 11 give the moment properties and asymptotic behavior of middle, intermediate and extreme order statistics. Chapters 12 to 15 discuss estimation of parameters and their properties. Chapters 16 to 20 deal with order statistics from extended samples, record values, characterizations, order statistics from F-alpha distribution and generalized order statistics. Chapter 21 contains several interesting problems with hints to solve them.

Summer research grant and sabbatical leave from Rider university enabled the first author to complete his part of the work. The work of the second author was supported by the grant RFFI 10-01-00314 and the grant of Science School 4472.2010. The third author is grateful to Miami Dade College for all the supports including STEM grants.

Contents

Chapter 1

Basic Definitions

We introduce the basic definitions. They are as follows:

$X_1, X_2, \ldots, X_n$ – *initial random variables.*

As a rule in the sequel we will suppose that these random variables are independent and have a common distribution function (d.f.) F. It enables us to consider the set $\{X_1, X_2, \ldots, X_n\}$ as a sample of size n taken from the population distribution F. The set of the observed values $\{x_1, x_2, \ldots, x_n\}$ of random variables $X_1, X_2, \ldots, X_n$ is called a realization of the sample. We can simply say also that $X_1, X_2, \ldots, X_n$ present n independent observations on X, where X is a random variable, having a d.f. F.

$X_{1,n} \leqslant X_{2,n} \leqslant \cdots \leqslant X_{n,n}$ *denotes variational series based on random variables $X_1, X_2, \ldots, X_n$. If X's are independent and identically distributed one can say that it is the variational series based on a sample $X_1, X_2, \ldots, X_n$. Elements $X_{k,n}$, $1 \leqslant k \leqslant n$, are called order statistics (order statistics based on a sample $X_1, X_2, \ldots, X_n$; order statistics from a d.f. F; ordered observations on X). We denote $x_{1,n}, x_{2,n}, \ldots, x_{n,n}$ as observed values of $X_{1,n}, X_{2,n}, \ldots, X_{n,n}$ and call them as a realization of order statistics. Let us note that $X_{1,n} = m(n) = \min\{X_1, X_2, \ldots, X_n\}$ and $X_{n,n} = M(n) = \max\{X_1, X_2, \ldots, X_n\}$, $n = 1, 2, \ldots$.*

$F_n^*(x) = \dfrac{1}{n} \sum_{k=1}^{n} 1_{\{X_k \leqslant x\}}$ *denotes the empirical (or sample) distribution function. Here $1_{\{X \leqslant x\}}$ is a random indicator, which equals 1, if $X \leqslant x$, and equals 0, if $X > x$.*

Let us mention that

$$
F_n^*(x) = \begin{cases} 0, & \text{if } x < X_{1,n}, \\ \dfrac{k}{n}, & \text{if } X_{k,n} \leqslant x < X_{k+1,n}, \ 1 \leqslant k \leqslant n-1, \text{ and} \\ 1, & \text{if } x \geqslant X_{n,n}. \end{cases}
$$

Together with a random sample $X_1, X_2, \ldots, X_n$ we can consider a vector of ranks

$$(R(1), R(2), \ldots, R(n)),$$

M. Ahsanullah et al., *An Introduction to Order Statistics*,
Atlantis Studies in Probability and Statistics 3,
DOI: 10.2991/978-94-91216-83-1_1, © Atlantis Press 2013

where

$$R(m) = \sum_{k=1}^{n} 1_{\{X_m \geqslant X_k\}}, \quad m = 1, 2, \ldots, n.$$

Ranks provide the following equalities for events:

$$\{R(m) = k\} = \{X_m = X_{k,n}\}, \quad m = 1, 2, \ldots, n, \quad k = 1, 2, \ldots, n.$$

Symmetrically antiranks $\Delta(1), \Delta(2), \ldots, \Delta(n)$ *are defined by equalities*

$$\{\Delta(k) = m\} = \{X_{k,n} = X_m\}, \quad m = 1, 2, \ldots, n, \quad k = 1, 2, \ldots, n.$$

One more type of ranks is presented by sequential ranks. For any sequence of random variables $X_1, X_2, \ldots$ *we introduce sequential ranks* $\rho(1), \rho(2), \ldots$ *as follows:*

$$\rho(m) = \sum_{k=1}^{m} 1_{\{X_m \geqslant X_k\}}, \quad m = 1, 2, \ldots.$$

Sequential rank $\rho(m)$ *shows a position of a new coming observation* X_m *among its predecessors* $X_1, X_2, \ldots, X_{m-1}$.

Sometimes we need to investigate ordered random variables. Indeed, we always can order a sequence of values. For example, if $a_1 = 3$, $a_2 = 1$, $a_3 = 3$, $a_4 = 2$ and $a_5 = 8$, then ordered values can be presented as 1, 2, 3, 3, 8 (non-decreasing order) or 8, 3, 3, 2, 1 (non-increasing order). Let us have now some random variables $X_1, X_2, \ldots, X_n$ defined on a common probability space $(\Omega, \mathfrak{I}, P)$. Each random variable maps Ω into $\mathfrak{R}$, the real line. It means that for any elementary event $\omega \in \Omega$ we have n real values $X_1(\omega), X_2(\omega), \ldots, X_n(\omega)$, which can be arranged in nondecreasing order. It enables us to introduce new random variables $X_{1,n} = X_{1,n}(\omega), X_{2,n} = X_{2,n}(\omega), \ldots, X_{n,n} = X_{n,n}(\omega)$ defined on the same probability space $(\Omega, \mathfrak{I}, P)$ as follows: for each $\omega \in \Omega$, $X_{1,n}(\omega)$ coincides with the smallest of the values $X_1(\omega), X_2(\omega), \ldots, X_n(\omega)$, $X_{2,n}(\omega)$ is the second smallest of these values, $X_{3,n}(\omega)$ is the third smallest, $\ldots$, and $X_{n,n}(\omega)$ is assigned the largest of the values $X_1(\omega), X_2(\omega), \ldots, X_n(\omega)$. Thus, a set of n arbitrary random variables $X_1, X_2, \ldots, X_n$ generates another set of random variables $X_{1,n}, X_{2,n}, \ldots, X_{n,n}$, already ordered. The used construction provides two important equalities:

$$P\{X_{1,n} \leqslant X_{2,n} \leqslant \cdots \leqslant X_{n,n}\} = 1 \tag{1.1}$$

and

$$X_{1,n} + X_{2,n} + \cdots + X_{n,n} = X_1 + X_2 + \cdots + X_n. \tag{1.2}$$

Exercise 1.1. Let a set of elementary events Ω consist of two elements ω_1 and ω_2 and random variables X_1 and X_2 be defined as follows: $X_1(\omega_1) = 0$, $X_1(\omega_2) = 3$, $X_2(\omega_1) = 1$, $X_2(\omega_2) = 2$. Describe ordered random variables $X_{1,2}$ and $X_{2,2}$ as functions on Ω.

Exercise 1.2. Let now $\Omega = [0,1]$ and three random variables X_1, X_2 and X_3 are defined as follows: $X_1(\omega) = \omega$, $X_2(\omega) = 1 - \omega$, $X_3(\omega) = 1/4$, $\omega \in [0,1]$. What is the structure of functions $X_{k,3}(\omega)$, $k = 1, 2, 3$?

Definition 1.1. We say that

$$X_{1,n} \leqslant X_{2,n} \leqslant \cdots \leqslant X_{n,n}$$

is the variational series based on random variables $X_1, X_2, \ldots, X_n$. Elements $X_{k,n}$, $k = 1, 2, \ldots, n$, of variational series are said to be order ststistics.

Very often in mathematical statistics we deal with sequences of independent random variables having a common d.f. F. Then a collection $X_1, X_2, \ldots, X_n$ can be interpreted as a random sample of size n from the d.f. F. We can say also that $X_1, X_2, \ldots, X_n$ present n independent observations on X, where X is a random variable, having a d.f. F. Hence in this situation we deal with the variational series and order statistics based on a sample X_1, $X_2, \ldots, X_n$. We can also say that $X_{k,n}$, $1 \leqslant k \leqslant n$, are order statistics from a d.f. F; or, for example, ordered observations on X. As a result of a statistical experiment we get a set of the observed values $\{x_1, x_2, \ldots, x_n\}$ of random variables $X_1, X_2, \ldots, X_n$. This set is called a realization of the sample. Analogously observed values of $X_{1,n}, X_{2,n}, \ldots, X_{n,n}$ we denote $x_{1,n}, x_{2,n}, \ldots, x_{n,n}$ and call a realization of order statistics.

In the sequel we will consider, in general, sequences of independent random variables. Here we must distinguish two important situations, namely, the case of continuous d.f.'s F and the case, when d.f.'s F have some discontinuity points.

The structure of order statistics essentially differs in these two situations. Let us try to show this difference.

Exercise 1.3. Let X_1 and X_2 be independent random variables with continuous d.f.'s F_1 and F_2. Prove that $P\{X_1 = X_2\} = 0$.

Exercise 1.4. Let

$$X_{1,n} \leqslant X_{2,n} \leqslant \cdots \leqslant X_{n,n}$$

be the variational series based on independent random variables $X_1, X_2, \ldots, X_n$ with continuous d.f.'s $F_1, F_2, \ldots, F_n$. Show that in this case

$$P\{X_{1,n} < X_{2,n} < \cdots < X_{n,n}\} = 1.$$

Exercise 1.5. Let $X_1, X_2, \ldots, X_n$ be independent random variables having the uniform distribution on the set $\{1, 2, 3, 4, 5, 6\}$. This situation corresponds, for instance, to the case then a die is rolled n times. Find

$$p_n = \{X_{1,n} < X_{2,n} < \cdots < X_{n,n}\}, \quad n = 1, 2, \ldots.$$

Exercise 1.6. Let $X_1, X_2, \ldots, X_n$ be n independent observations on random variable X, having the geometric distribution, that is

$$P\{X = m\} = (1 - p)p^m, \quad m = 0, 1, 2, \ldots,$$

and $X_{k,n}$ be the corresponding order statistics. Find

$$p_n = P\{X_{1,n} < X_{2,n} < \cdots < X_{n,n}\}, \quad n = 2, 3, \ldots.$$

Unless otherwise is proposed, in the sequel we suppose that X's are independent random variables having a common continuous d.f. F. In this situation order statistics satisfy inequalities

$$X_{1,n} < X_{2,n} < \cdots < X_{n,n}$$

with probability one.

There are different types of rank statistics, which help us to investigate ordered random variables. Together with a sample $X_1, X_2, \ldots, X_n$ we will consider a vector of ranks

$$(R(1), R(2), \ldots, R(n)),$$

elements of which show the location of X's in the variational series

$$X_{1,n} \leqslant X_{2,n} \leqslant \cdots \leqslant X_{n,n}.$$

Definition 1.2. Random variables $R(1), \ldots, R(n)$ given by equalities

$$R(m) = \sum_{k=1}^{n} 1_{\{X_m \geqslant X_k\}}, \quad m = 1, 2, \ldots, n. \tag{1.3}$$

are said to be ranks corresponding to the sample $X_1, X_2, \ldots, X_n$.

Since we consider the situation, when different X's can coincide with zero probability, (1.3) can be rewritten in the following form:

$$R(m) = 1 + \sum_{k=1}^{n} 1_{\{X_m > X_k\}}. \tag{1.4}$$

The following equality for events is a corollary of (1.4):

$$\{R(m) = k\} = \{X_{k,n} = X_m\}, \quad m = 1, 2, \ldots, n, \quad k = 1, 2, \ldots, n. \tag{1.5}$$

Another form of (1.5) is

$$X_m = X_{R(m),n}, \quad m = 1, 2, \ldots, n. \tag{1.6}$$

Exercise 1.7. For any $m = 1, 2, \ldots, n$, prove that $R(m)$ has the discrete uniform distribution on set $\{1, 2, \ldots, n\}$.

We know that i.i.d. random variables $X_1, X_2, \ldots, X_n$ taken from a continuous distribution have no coincidences with probability one. Hence, realizations $(r(1), \ldots, r(n))$ of the corresponding vector of ranks $(R(1), R(2), \ldots, R(n))$ represent all permutations of values $1, 2, \ldots, n$. Any realization $(r(1), \ldots, r(n))$ corresponds to the event

$$\left(X_{\delta(1)} < X_{\delta(2)} < \cdots < X_{\delta(n)} \right),$$

where $\delta(r(k)) = k$. For example, the event

$$\{R(1) = n, \ R(2) = n - 1, \ldots, \ R(n) = 1\}$$

is equivalent to the event

$$\{X_n < X_{n-1} < \cdots < X_1\}.$$

Taking into account the symmetry of the sample $X_1, X_2, \ldots, X_n$, we obtain that events

$$\left(X_{\delta(1)} < X_{\delta(2)} < \cdots < X_{\delta(n)} \right)$$

have the same probabilities for any permutations $(\delta(1), \ldots, \delta(n))$ of numbers $1, 2, \ldots, n$. Hence

$$P\{R(1) = r(1), \ R(2) = r(2), \ldots, \ R(n) = r(n)\} =$$

$$P\left\{ \left(X_{\delta(1)} < X_{\delta(2)} < \cdots < X_{\delta(n)} \right) \right\} = \frac{1}{n!} \tag{1.7}$$

for any permutation $(r(1), \ldots, r(n))$ of numbers $1, 2, \ldots, n$.

Exercise 1.8. For fixed n and $k < n$, find

$$P\{R(1) = r(1), \ R(2) = r(2), \ldots, \ R(k) = r(k)\},$$

where $r(1), \ r(2), \ldots, \ r(k)$ are different numbers taken from the set $\{1, 2, \ldots, n\}$.

Exercise 1.9. Show that ranks $R(1), \ R(2), \ldots, \ R(n)$ are dependent random variables for any $n = 2, 3, \ldots$.

The dependence of ranks is also approved by the evident equality

$$R(1) + R(2) + \cdots + R(n) = 1 + 2 + \cdots + n = \frac{n(n+1)}{2}, \tag{1.8}$$

which is valid with probability one always, when X's are independent and have a common continuous distribution.

It follows from (1.7) that ranks are exchangeable random variables: for any permutation $(\alpha(1), \alpha(2), \ldots, \alpha(n))$ of numbers $1, 2, \ldots, n$, vectors $(R(\alpha(1)), \ldots, R(\alpha(n)))$ and $(R(1), \ldots, R(n))$ have the same distributions.

Exercise 1.10. Find expectations and variances of $R(k)$, $1 \leqslant k \leqslant n$.

Exercise 1.11. Find the covariance $\mathrm{Cov}(R(k), R(m))$ and the correlation coefficients $\rho(R(k), R(m))$ between $R(k)$ and $R(m)$, $1 \leqslant k, m \leqslant n$.

Exercise 1.12. Find $ER(1)R(2)\cdots R(n-1)$ and $ER(1)R(2)\cdots R(n)$.

Above we mentioned that any realization $(r(1), \ldots, r(n))$ of $(R(1), R(2), \ldots, R(n))$ corresponds to the event

$$(X_{\delta(1)} < X_{\delta(2)} < \cdots < X_{\delta(n)}),$$

where $\delta(r(k)) = k$.

Here $\delta(k)$ denotes the index of X, the rank of which for this realization takes on the value k. For different realizations of the vector $(R(1), R(2), \ldots, R(n))$, $\delta(k)$ can take on different values from the set $\{1, 2, \ldots, n\}$ and we really deal with new random variables, which realizations are $\delta(1)$, $\delta(2), \ldots, \delta(n)$.

Definition 1.3. Let $X_1, X_2, \ldots, X_n$ be a random sample of size n taken from a continuous distribution and $X_{1,n}, X_{2,n}, \ldots, X_{n,n}$ be the corresponding order statistics. Random variables $\Delta(1), \Delta(2), \ldots, \Delta(n)$, which satisfy the following equalities:

$$\{\Delta(m) = k\} = \{X_{m,n} = X_k\}, \quad m = 1, 2, \ldots, n, \quad k = 1, 2, \ldots, n, \tag{1.9}$$

are said to be antiranks.

The same arguments, which we used for ranks, show that any realization $(\delta(1), \delta(2), \ldots, \delta(n))$ of the vector $(\Delta(1), \Delta(2), \ldots, \Delta(n))$ is a permutation of numbers $(1, 2, \ldots, n)$ and all $n!$ realizations have equal probabilities, $1/n!$ each. Indeed, vectors of antiranks are tied closely with the corresponding order statistics and vectors of ranks. In fact, for any k and m equalities

$$\{\Delta(k) = m\} = \{X_{k,n} = X_m\} = \{R(m) = k\} \tag{1.10}$$

hold with probability one. We can write also the following identities for ranks and antiranks:

$$\Delta(R(m)) = m \tag{1.11}$$

and

$$R(\Delta(m)) = m, \qquad (1.12)$$

which hold with probability one for any $m = 1, 2, \ldots, n$.

Exercise 1.13. Find the joint distribution of $\Delta(1)$ and $R(1)$.

While ranks and antiranks are associated with some random sample and its size n, there are rank statistics, which characterize a sequence of random variables $X_1, X_2, \ldots$.

Definition 1.4. Let $X_1, X_2, \ldots, X_n$ be independent random variables, having continuous (not necessary identical) distributions. Random variables $\rho(1), \rho(2), \ldots$ given by equalities:

$$\rho(m) = \sum_{k=1}^{m} 1_{\{X_m \geqslant X_k\}}, \quad m = 1, 2, \ldots. \qquad (1.13)$$

are said to be sequential ranks.

Sequential ranks are not associated with some sample of a fixed size n. In fact, $\rho(m)$ shows a position of a new coming observation X_m among its predecessors $X_1, X_2, \ldots, X_{m-1}$. For instance, if $\rho(m) = 1$, then X_m is less then $X_{1,m-1}$ and it means that

$$X_m = X_{1,m}.$$

In general, $\rho(m) = k$ implies that

$$X_m = X_{k,m}.$$

It is not difficult to see that $\rho(m)$ takes on the values $1, 2, \ldots, m$. If independent random variables $X_1, X_2, \ldots, X_m$ have the same continuous distribution then the standard arguments used above enable us to see that for any $m = 1, 2, \ldots$,

$$P\{\rho(m) = k\} = P\{X_m = X_{k,m}\} = 1/m, \quad k = 1, 2, \ldots, m. \qquad (1.14)$$

Exercise 1.14. Let $X_1, X_2, \ldots$ be independent random variables with a common continuous d.f. F. Prove that the corresponding sequential ranks $\rho(1), \rho(2), \ldots$ are independent.

In example 1.1 we deal with different types of ranks.

Example 1.1. Let the following data represent the lifetimes (hours) of 15 batteries (realization of some sample of size 15):

$$x_1 = 20.3 \quad x_2 = 17.2 \quad x_3 = 15.4 \quad x_4 = 16.8 \quad x_5 = 24.1$$
$$x_6 = 12.6 \quad x_7 = 15.0 \quad x_8 = 18.1 \quad x_9 = 19.1 \quad x_{10} = 21.3$$
$$x_{11} = 22.3 \quad x_{12} = 16.4 \quad x_{13} = 13.5 \quad x_{14} = 25.8 \quad x_{15} = 16.9$$

Being ordered these observations give us realizations of order statistics:

$$x_{1,15} = 12.6 \quad x_{2,15} = 13.5 \quad x_{3,15} = 15.0 \quad x_{4,15} = 15.4 \quad x_{5,15} = 16.4$$
$$x_{6,15} = 16.8 \quad x_{7,15} = 16.9 \quad x_{8,15} = 17.2 \quad x_{9,15} = 18.1 \quad x_{10,15} = 19.1$$
$$x_{11,15} = 20.3 \quad x_{12,15} = 21.3 \quad x_{13,15} = 22.3 \quad x_{14,15} = 24.1 \quad x_{15,15} = 25.8$$

Realizations of ranks are given as follows:

$$r(1) = 11, \quad r(2) = 8, \quad r(3) = 4, \quad r(4) = 6, \quad r(5) = 14,$$
$$r(6) = 1, \quad r(7) = 3, \quad r(8) = 9, \quad r(9) = 10, \quad r(10) = 12,$$
$$r(11) = 13, \quad r(12) = 5, \quad r(13) = 2, \quad r(14) = 15, \quad r(15) = 7.$$

Antiranks are presented by the sequence:

$$\delta(1) = 6, \quad \delta(2) = 13, \quad \delta(3) = 7, \quad \delta(4) = 3, \quad \delta(5) = 12,$$
$$\delta(6) = 4, \quad \delta(7) = 15, \quad \delta(8) = 2, \quad \delta(9) = 8, \quad \delta(10) = 9,$$
$$\delta(11) = 1, \quad \delta(12) = 10, \quad \delta(13) = 11, \quad \delta(14) = 5, \quad \delta(15) = 14.$$

The sequential ranks are:

$$1; \; 1; \; 1; \; 2; \; 5; \; 1; \; 2; \; 6; \; 7; \; 9; \; 10; \; 4; \; 2; \; 14; \; 7$$

Very often to estimate an unknown population distribution function F a statistician uses the so-called empirical (sample) distribution function

$$F_n^*(x) = \frac{1}{n} \sum_{k=1}^{n} 1_{\{X_k \leqslant x\}}. \tag{1.15}$$

Empirical distribution functions are closely tied with order statistics so far as

$$F_n^*(x) = \begin{cases} 0, & \text{if } x < X_{1,n}, \\ 1, & \text{if } x \geqslant X_{n,n}, \text{ and} \\ \dfrac{k}{n}, & \text{if } X_{k,n} \leqslant x < X_{k+1,n}, 1 \leqslant k \leqslant n-1. \end{cases} \tag{1.16}$$

Exercise 1.15. Find the expectation and the variance of $F_n^*(x)$.

Check your solutions

Exercise 1.1 (solution). On comparing values of X_1 and X_2 one can see that

$$X_{1,2}(\omega_1) = 0, \quad X_{1,2}(\omega_2) = 2 \text{ and } X_{2,2}(\omega_1) = 1, \quad X_{2,2}(\omega_2) = 3.$$

Thus, $X_{1,2}$ partially coincides with $X_1(\omega_1)$ and partially with X_2 (on ω_2) as well as $X_{1,2}$ which coincides with $X_1(\omega)$ and X_2.

Exercise 1.2 (answers).

$$X_{1,3}(\omega) = \begin{cases} X_1(\omega) = \omega, & \text{if } 0 \leqslant \omega \leqslant 1/4; \\ X_3(\omega) = 1/4, & \text{if } 1/4 < x < 3/4, \\ X_2(\omega), & \text{if } 3/4 \leqslant x \leqslant 1; \end{cases}$$

$$X_{2,3}(\omega) = \begin{cases} X_3(\omega) = 1/4, & \text{if } 0 \leqslant \omega \leqslant 1/4 \text{ and if } 3/4 \leqslant \omega \leqslant 1, \\ X_2(\omega) = \omega, & \text{if } 1/4 < \omega \leqslant 1/2, \\ X_3(\omega) = \omega, & \text{if } 1/2 < \omega < 3/4; \end{cases}$$

$$X_{3,3}(\omega) = \begin{cases} X_3(\omega) = 1 - \omega, & \text{if } 0 \leqslant \omega \leqslant 1/2, \\ X_1(\omega) = \omega, & \text{if } 1/2 < x \leqslant 1. \end{cases}$$

Exercise 1.3 (solution). Since X_1 has a continuous d.f. we have that

$$P\{X_1 = x\} = F_1(x) - F_1(x - 0) = 0 \text{ for any } x.$$

Then, taking into account the independence of X_1 and X_2 we obtain that

$$P\{X_1 = X_2\} = \int_{-\infty}^{\infty} P\{X_1 = X_2 \mid X_2 = x\} dF_2(x)$$

$$= \int_{-\infty}^{\infty} P\{X_1 = x \mid X_2 = x\} dF_2(x)$$

$$= \int_{-\infty}^{\infty} P\{X_1 = x\} dF_2(x) = 0.$$

Exercise 1.4 (solution). Let A be an event such that there are at least two coincidences among $X_1, X_2, \ldots, X_n$ and $A_{ij} = \{X_i = X_j\}$.

We know from Exercise 1.3 that $P\{A_{ij}\} = 0$ if $i \neq j$.

The assertion of exercise 1.4 holds since

$$1 - P\{X_{1,n} < X_{2,n} < \cdots < X_{n,n}\} = P\{A\} \leqslant \sum_{i \neq j} P\{A_{ij}\} = 0.$$

Exercise 1.5 (solution). It is evident that $p_1 = 1$ and $p_n = 0$, if $n > 6$. One can see that p_n is equal to the probability that the die shows n different values. Hence $p_n = 6!/(6-n)!6^n$, $n = 2, \ldots, 6$, and, in particular, $p_2 = 5/6$, $p_3 = 5/9$, $p_4 = 5/18$, $p_5 = 5/54$ and $p_6 = 5/324$.

Exercise 1.6 (solution). In this case

$$p_n = n! P\{X_1 < X_2 < \cdots < X_n\} = n! \sum_{k_1=0}^{\infty} (1-p) p^{k_1} \sum_{k_2=k_1+1}^{\infty} (1-p) p^{k_2} \cdots \sum_{k_n=k_{n-1}+1}^{\infty} (1-p) p^{k_n}.$$

Sequential simplifying of this series gives us the following expression for p_n:

$$p_n = \frac{n!(1-p)^n p^{n(n-1)/2}}{\prod_{k=1}^{n}(1-p)^k} = \frac{n! p^{n(n-1)/2}}{(1+p)(1+p+p^2)\cdots(1+p+\cdots+p^{n-1})}.$$

In particular,

$$p_2 = \frac{2p}{1+p} \quad \text{and} \quad p_3 = \frac{6p^3}{(1+p)(1+p+p^2)}.$$

Exercise 1.7 (solution). a) The symmetry argument enables us to prove the necessary statement for one rank only, say for $R(1)$. From (1.4), on using the total probability rule and independence of X's, one has that

$$P\{R(1) = k\} = P\left\{ \sum_{s=2}^{n} 1_{\{X_1 > X_s\}} = k - 1 \right\}$$

$$= \int_{-\infty}^{\infty} P\left\{ \sum_{s=2}^{n} 1_{\{X_1 > X_s\}} = k - 1 \mid X_1 = x \right\} dF(x)$$

$$= \int_{-\infty}^{\infty} P\left\{ \sum_{s=2}^{n} 1_{\{X_s < x\}} = k - 1 \mid X_1 = x \right\} dF(x)$$

$$= \int_{-\infty}^{\infty} P\left\{ \sum_{s=2}^{n} 1_{\{X_s < x\}} = k - 1 \right\} dF(x).$$

Indeed, the sum under probability sign has the binomial distribution with parameters $n - 1$ and $F(x)$ and hence we finally have that

$$P\left\{ \sum_{s=2}^{n} 1_{\{X_s < x\}} = k - 1 \right\} = \binom{n-1}{k-1} (F(x))^{k-1} (1 - F(x))^{n-k}$$

and

$$\int_{-\infty}^{\infty} P\left\{ \sum_{s=2}^{n} 1_{\{X_s < x\}} = k - 1 \right\} dF(x) = \binom{n-1}{k-1} \int_{-\infty}^{\infty} (F(x))^{k-1} (1 - F(x))^{n-k} dF(x)$$

$$= \binom{n-1}{k-1} \int_{0}^{1} u^{k-1} (1-u)^{n-k} du = \binom{n-1}{k-1} B(k, n-k+1)$$

$$= \binom{n-1}{k-1} \frac{(k-1)!(n-k)!}{n!} = \frac{1}{n}.$$

Above we used the following formula for Beta function:

$$B(m,n) = \int_{0}^{1} u^{m-1} (1-u)^{n-1} du = \frac{(m-1)!(n-1)!}{(m+n-1)!}.$$

b) A less rigorous proof also is based on symmetry argument. We can note that for any $k = 1, 2, \ldots, n$, events $\{X_{k,n} = X_1\}, \{X_{k,n} = X_2\}, \ldots, \{X_{k,n} = X_n\}$ must have equal probabilities. Their sum equals one. Hence $P\{X_{k,n} = X_m\} = 1/n$ for any $1 \leqslant m, k \leqslant n$.

Exercise 1.8 (solution). The event $\{R(1) = r(1), \ldots, R(k) = r(k)\}$ is the union of $(n-k)!$ events

$$\{R(1) = r(1), \ldots, R(k) = r(k), R(k+1) = s(1), \ldots, R(n) = s(n-k)\},$$

where $(s(1), s(2), \ldots, s(n-k))$ are permutations of $(n-k)$ numbers taken from the set $\{1, 2, \ldots, n\} \smallsetminus \{r(1), r(2), \ldots, r(k)\}$.

Since each of events

$$\{R(1) = r(1), \ldots, R(k) = r(k), \ R(k+1) = s(1), \ldots, \ R(n) = s(n-k)\}$$

has probability $1/n!$, we get that

$$P\{R(1) = r(1), \ldots, \ R(k) = r(k)\} = \frac{(n-k)!}{n!}.$$

Exercise 1.9 (solution). On comparing equalities

$$P\{R(1) = r(1), R(2) = r(2), \ldots, R(n) = r(n)\} = \frac{1}{n!} \ \text{ and } \ P\{R(k) = r(k)\} = \frac{1}{n},$$

we obtain that

$$\frac{1}{n!} = P\{R(1) = r(1), R(2) = r(2), \ldots, R(n) = r(n)\}$$

$$\neq P\{R(1) = r(1)\} \cdots P\{R(n) = r(n)\} = \frac{1}{n^n}.$$

It means that ranks $R(1), \ R(2), \ldots, \ R(n)$ are dependent.

Exercise 1.10 (answers). $\mathrm{E}R(k) = \dfrac{n+1}{2}$, $\mathrm{Var}\,R(k) = \dfrac{(n-1)^2}{12}$, $1 \leqslant k \leqslant n$.

Exercise 1.11 (solution). Indeed,

$$\mathrm{Cov}\,(R(k), R(k)) = \mathrm{Var}\,R(k) = \frac{(n-1)^2}{12}, \quad k = 1, 2, \ldots, n,$$

as it follows from Exercise 1.10, and $\rho(R(k), R(k)) = 1$ for any k. Exchangeability of ranks implies that

$$\mathrm{Cov}\,(R(k), R(m)) = \mathrm{cov}\,(R(1), R(2))$$

and $\rho(R(k), R(m)) = \rho(R(1), R(2))$ for any $1 \leqslant k \neq m \leqslant n$. Then, we obtain from (1.8) that

$$0 = \mathrm{Var}\,(R(1) + R(2) + \cdots + R(n))$$

$$= \sum_{k=1}^{n} \mathrm{Var}\,R(k) + 2 \sum_{1 \leqslant k < m \leqslant n} \mathrm{Cov}\,(R(k), R(m))$$

$$= n\,\mathrm{Var}\,R(1) + 2\binom{n}{2}\mathrm{Cov}\,(R(1), R(2)).$$

Hence,

$$\rho(R(1), R(2)) = \frac{\mathrm{cov}\,(R(1), R(2))}{(\mathrm{Var}\,R(1))^{1/2}(\mathrm{Var}\,R(2))^{1/2}} = \frac{\mathrm{cov}\,(R(1), R(2))}{\mathrm{Var}\,R(1)} = -\frac{n}{2}\binom{n}{2} = -\frac{1}{n-1}$$

and

$$\mathrm{cov}\,(R(1), R(2)) = -n\frac{\mathrm{Var}\,R(1)}{n(n-1)} = -\frac{n+1}{12}.$$

Exercise 1.12 (solution). Since $(R(1), R(2), \ldots, R(n))$ is a random permutation of numbers $1, 2, \ldots, n$, we have that $R(1)R(2) \cdots R(n) = n!$ with probability one. It implies that

$$E R(1)R(2) \cdots R(n) = n!.$$

Analogously we can write that

$$E R(1)R(2) \cdots R(n-1) = E\left(\frac{n!}{R(n)}\right) = n! E\left(\frac{1}{R(n)}\right).$$

Since $R(n)$ takes on the values $1, 2, \ldots, n$ with probabilities $1/n$, we have now that

$$E\left(\frac{1}{R(n)}\right) = \frac{1}{n}\sum_{k=1}^{n}\frac{1}{k}$$

and

$$E R(1)R(2) \cdots R(n-1) = (n-1)!\sum_{k=1}^{n}\frac{1}{k}.$$

Exercise 1.13 (answers).

$$p(m,s) = \begin{cases} P\{\Delta(1) = m,\ R(1) = s\} = \dfrac{1}{n}, & \text{if } m = s = 1; \\[2mm] \dfrac{1}{n(n-1)}, & \text{if } s \neq 1,\ m \neq 1,\ \text{and} \\[2mm] 0, & \text{otherwise.} \end{cases}$$

Exercise 1.14 (solution). Since $P\{\rho(m) = k\} = 1/m$, $k = 1, 2, \ldots, m$, it needs to show that for any $n = 1, 2, \ldots$, and any $a(k)$, taking on values $1, 2, \ldots, k$, $1 \leqslant k \leqslant n$,

$$P\{\rho(1) = a(1),\ \rho(2) = a(2), \ldots, \rho(n) = a(n)\}$$

$$= P\{\rho(1) = a(1)\}P\{\rho(2) = a(2)\} \cdots P\{\rho(n) = a(n)\} = \frac{1}{n!}.$$

Fix n and consider ranks $R(1), R(2), \ldots, R(n)$. It is not difficult to see that a set $\{a(1), a(2), \ldots, a(n)\}$ uniquely determines values $r(1), r(2), \ldots, r(n)$ of $R(1), R(2), \ldots, R(n)$. In fact, $r(n) = a(n)$. Further, $r(n-1) = a(n-1)$, if $a(n) > a(n-1)$, and $r(n-1) = a(n-1) + 1$, if $a(n) \leqslant a(n-1)$. The value of $R(n-2)$ is analogously determined by values $a(n)$, $a(n-1)$ and $a(n-2)$ and so on. Hence, each of $n!$ events

$$\{\rho(1) = a(1), \rho(2) = a(2), \ldots, \rho(n) = a(n)\}$$

coincides with one of $n!$ events

$$\{R(1) = r(1),\ R(2) = r(2), \ldots,\ R(n) = r(n)\}.$$

For instance,

$$\{\rho(1) = 1,\ \rho(2) = 1, \ldots,\ \rho(n) = 1\} = \{R(1) = n,\ R(2) = n-1, \ldots,\ R(n) = 1\}.$$

Since

$$P\{R(1) = r(1),\ R(2) = r(2),\ldots,\ R(n) = r(n)\} = \frac{1}{n!}$$

for any permutation $(r(1), r(2),\ldots, r(n))$ of the numbers $1, 2,\ldots, n$, we have that

$$P\{\rho(1) = a(1)\}P\{\rho(2) = a(2)\}\cdots P\{\rho(n) = a(n)\} = \frac{1}{n!}$$

for any set $\{a(1), a(2),\ldots, a(n)\}$, where $1 \leqslant a(k) \leqslant k$, $k = 1, 2,\ldots, n$.

Exercise 1.15 (answers).

$$E F_n^*(x) = F(x),$$

$$\operatorname{Var} F_n^*(x) = \frac{F(x)(1 - F(x))}{n}.$$

Chapter 2

Distributions of Order Statistics

We give some important formulae for distributions of order statistics. For example,

$$F_{k:n}(x) = P\{X_{k,n} \leqslant x\} = I_{F(x)} \; (k, \, n-k+1),$$

where

$$I_x(a,b) = \frac{1}{B(a,b)} \int_0^x t^{a-1}(1-t)^{b-1} dt$$

denotes the incomplete Beta function. The probability density function of $X_{k,n}$ is given as follows:

$$f_{k:n}(x) = \frac{n!}{(k-1)!(n-k)!}(F(x))^{k-1}(1-F(x))^{n-k} f(x),$$

where f is a population density function. The joint density function of order statistics $X_{1,n}, X_{2,n}, \ldots, X_{n,n}$ has the form

$$f_{1,2,\ldots,n:n}(x_1, x_2, \ldots, x_n) = \begin{cases} n! \prod\limits_{k=1}^{n} f(x_k), & -\infty < x_1 < x_2 < \cdots < x_n < \infty, \text{ and} \\ 0, & \text{otherwise.} \end{cases}$$

There are some simple formulae for distribution of maxima and minima.

Example 2.1. Let random variables $X_1, X_2, \ldots, X_n$ have a joint d.f.

$$H(x_1, x_2, \ldots, x_n) = P\{X_1 \leqslant x_1, \, X_2 \leqslant x_2, \ldots, \, X_n \leqslant x_n\}.$$

Then d.f. of $M(n) = \max\{X_1, X_2, \ldots, X_n\}$ has the form

$$P\{M(n) \leqslant x\} = P\{X_1 \leqslant x, \, X_2 \leqslant x, \ldots, \, X_n \leqslant x\} = H(x, x, \ldots, x). \tag{2.1}$$

Similarly we can get the distribution of $m(n) = \min\{X_1, X_2, \ldots, X_n\}$.
One has

$$P\{m(n) \leqslant x\} = 1 - P\{m(n) > x\} = 1 - P\{X_1 > x, \, X_2 > x, \ldots, \, X_n > x\}. \tag{2.2}$$

M. Ahsanullah et al., *An Introduction to Order Statistics*,
Atlantis Studies in Probability and Statistics 3,
DOI: 10.2991/978-94-91216-83-1_2, © Atlantis Press 2013

Exercise 2.1. Find the joint distribution function of $M(n-1)$ and $M(n)$.

Exercise 2.2. Express d.f. of $Y = \min\{X_1, X_2\}$ in terms of joint d.f.

$$H(x_1, x_2) = P\{X_1 \leqslant x_1, X_2 \leqslant x_2\}.$$

Exercise 2.3. Let $H(x_1, x_2)$ be the joint d.f. of X_1 and X_2. Find the joint d.f. of

$$Y = \min\{X_1, X_2\} \quad \text{and} \quad Z = \max\{X_1, X_2\}.$$

From (2.1) and (2.2) one obtains the following elementary expressions for the case, when $X_1, X_2, \ldots, X_n$ present a sample from a population d.f. F (not necessary continuous):

$$P\{M(n) \leqslant x\} = F^n(x) \tag{2.3}$$

and

$$P\{m(n) \leqslant x\} = 1 - (1 - F(x))^n. \tag{2.4}$$

Exercise 2.4. Let $X_1, X_2, \ldots, X_n$ be a sample of size n from a geometrically distributed random variable X, such that $P\{X = m\} = (1-p)p^m$, $m = 0, 1, 2, \ldots$.
Find

$$P\{Y \geqslant r, Z < s\}, \quad r < s,$$

where $Y = \min\{X_1, X_2, \ldots, X_n\}$ and $Z = \max\{X_1, X_2, \ldots, X_n\}$.

There is no difficulty to obtain d.f.'s for single order statistics $X_{k,n}$. Let

$$F_{k,n}(x) = P\{X_{k,n} \leqslant x\}.$$

We see immediately from (2.3) and (2.4) that $F_{n,n}(x) = F^n(x)$ and $F_{1:n}(x) = 1 - (1 - F(x))^n$.
The general formula for $F_{k:n}(x)$ is not much more complicated. In fact,

$$F_{k:n}(x) = P\{ \text{ at least } k \text{ variables among } X_1, X_2, \ldots, X_n \text{ are less or equal } x\}$$

$$= \sum_{m=k}^{n} P\{ \text{ exactly } m \text{ variables among } X_1, X_2, \ldots, X_n \text{ are less or equal } x\}$$

$$= \sum_{m=k}^{n} \binom{n}{m} (F(x))^m (1 - F(x))^{n-m}, \quad 1 \leqslant k \leqslant n. \tag{2.5}$$

Exercise 2.5. Prove that identity

$$\sum_{m=k}^{n} \binom{n}{m} x^m (1-x)^{n-m} = I_x(k, n-k+1) \tag{2.6}$$

holds for any $0 \leqslant x \leqslant 1$, where

$$I_x(a,b) = \frac{1}{B(a,b)} \int_0^x t^{a-1}(1-t)^{b-1} dt \tag{2.7}$$

is the incomplete Beta function with parameters a and b, $B(a,b)$ being the classical Beta function.

By comparing (2.5) and (2.6) one obtains that

$$F_{k:n}(x) = I_{F(x)}(k, n-k+1). \tag{2.8}$$

Remark 2.1. It follows from (2.8) that $X_{k,n}$ has the beta distribution with parameters k and $n-k+1$, if X has the uniform on $[0,1]$ distribution.

Remark 2.2. Equality (2.8) is valid for any distribution function f.

Remark 2.3. If we have tables of the function $I_x(k, n-k+1)$, it is possible to obtain d.f. $F_{k:n}(x)$ for arbitrary d.f. F.

Exercise 2.6. Find the joint distribution of two order statistics $X_{r,n}$ and $X_{s,n}$.

Example 2.2. Let us try to find the joint distribution of all elements of the variational series $X_{1,n}, X_{2,n}, \ldots, X_{n,n}$.
It seems that the joint d.f.

$$F_{1,2,\ldots,n:n}(x_1, x_2, \ldots, x_n) = P\{X_{1,n} \leqslant x_1, X_{2,n} \leqslant x_2, \ldots, X_{n,n} \leqslant x_n\}$$

promises to be very complicated. Hence, we consider probabilities

$$P(y_1, x_1, y_2, x_2, \ldots, y_n, x_n) = P\{y_1 < X_{1,n} \leqslant x_1, y_2 < X_{2,n} \leqslant x_2, \ldots, y_n < X_{n,n} \leqslant x_n\}$$

for any values $-\infty \leqslant y_1 < x_1 \leqslant y_2 < x_2 \leqslant \cdots \leqslant y_n < x_n \leqslant \infty$.
It is evident, that the event

$$A = \{y_1 < X_{1,n} \leqslant x_1, y_2 < X_{2,n} \leqslant x_2, \ldots, y_n < X_{n,n} \leqslant x_n\}$$

is a union of $n!$ disjoint events
$A(\alpha(1), \alpha(2), \ldots, \alpha(n)) = \{y_1 < X_{\alpha(1)} \leqslant x_1, y_2 < X_{\alpha(2)} \leqslant x_2, \ldots, y_n < X_{\alpha(n)} \leqslant x_n\}$, where the vector $(\alpha(1), \alpha(2), \ldots, \alpha(n))$ runs all permutations of numbers $1, 2, \ldots, n$. The symmetry argument shows that all events $A(\alpha(1), \alpha(2), \ldots, \alpha(n))$ have the same probability. Note that this probability is equal to

$$\prod_{k=1}^{n} (F(x_k) - F(y_k)).$$

Finally, we obtain that

$$P\{y_1 < X_{1,n} \leqslant x_1, y_2 < X_{2,n} \leqslant x_2, \ldots, y_n < X_{n,n} \leqslant x_n\} = n! \prod_{k=1}^{n} (F(x_k) - F(y_k)). \tag{2.9}$$

Let now consider the case when our population distribution has a density function f. It means that for almost all X (i.e., except, possibly, a set of zero Lebesgue measure) $F'(x) = f(x)$. In this situation (2.9) enables us to find an expression for the joint probability density function (p.d.f.) (denote it $f_{1,2,\ldots,n}(x_1, x_2, \ldots, x_n)$) of order statistics $X_{1,n}, X_{2,n}, \ldots, X_{n,n}$. In fact, differentiating both sides of (2.9) with respect to $x_1, x_2, \ldots, x_n$, we get the important equality

$$f_{1,2,\ldots,n:n}(x_1, x_2, \ldots, x_n) = n! \prod_{k=1}^{n} f(x_k), \quad -\infty < x_1 < x_2 < \cdots < x_n < \infty. \tag{2.10}$$

Otherwise (if inequalities $x_1 < x_2 < \cdots < x_n$ fail) we naturally put

$$f_{1,2,\ldots,n:n}(x_1, x_2, \ldots, x_n) = 0.$$

Taking (2.10) as a starting point one can obtain different results for joint distributions of arbitrary sets of order statistics.

Example 2.3. Let $f_{m:n}$ denote the p.d.f. of $X_{m,n}$. We get from (2.10) that

$$f_{m:n}(x) = \int \cdots \int f_{1,2,\ldots,n:n}(x_1, \ldots, x_{k-1}, x, x_{k+1}, \ldots, x_n) dx_1 \cdots dx_{k-1} dx_{k+1} \cdots dx_n$$

$$= n! f(x) \int \cdots \int \prod_{k=1}^{m-1} f(x_k) \prod_{k=m+1}^{n} f(x_k) dx_1 \cdots dx_{m-1} dx_{m+1} \cdots dx_n, \tag{2.11}$$

where the integration is over the domain

$$-\infty < x_1 < \cdots < x_{m-1} < x < x_{m+1} < \cdots < x_n < \infty.$$

The symmetry of

$$\prod_{k=1}^{m-1} f(x_k)$$

with respect to $x_1, \ldots, x_{m-1}$, as well as the symmetry of

$$\prod_{k=m+1}^{n} f(x_k)$$

with respect to $x_{m+1}, \ldots, x_n$, helps us to evaluate the integral on the RHS of (2.11) as follows:

$$\int \cdots \int \prod_{k=1}^{m-1} f(x_k) \prod_{k=m+1}^{n} f(x_k) dx_1 \cdots dx_{m-1} dx_{m+1} \cdots dx_n$$

$$= \frac{1}{(m-1)!} \prod_{k=1}^{m-1} \int_{-\infty}^{x} f(x_k) dx_k \frac{1}{(n-m)!} \prod_{k=m+1}^{n} \int_{x}^{\infty} f(x_k) dx_k$$

$$= \frac{(F(x))^{m-1}(1-F(x))^{n-m}}{(m-1)!(n-m)!}. \tag{2.12}$$

Combining (2.11) and (2.12), we get that

$$f_{m:n}(x) = \frac{n!}{(m-1)!(n-m)!}(F(x))^{m-1}(1-F(x))^{n-m}f(x). \tag{2.13}$$

Indeed, equality (2.13) is immediately follows from the corresponding formula for d.f.'s of single order statistics (see (2.8), for example), but the technique, which we used to prove (2.13), is applicable for more complicated situations. The following exercise can illustrate this statement.

Exercise 2.7. Find the joint p.d.f.

$$f_{k(1),k(2),\ldots,k(r):n}(x_1,x_2,\ldots,x_r)$$

of order statistics $X_{k(1),n}, X_{k(2),n},\ldots, X_{k(r),n}$, where $1 \leqslant k(1) < k(2) < \cdots < k(r) \leqslant n$.

Remark 2.4. In the sequel we will often use the particular case of the joint probability density functions from exercise 2.7, which corresponds to the case $r = 2$. It turns out that

$$f_{i,j:n}(x_1,x_2) = \frac{n!}{(i-1)!(j-i-1)!(n-j)!}$$
$$\times (F(x_1))^{i-1}(F(x_2) - F(x_1))^{j-i-1}(1-F(x_2))^{n-j}f(x_1)f(x_2), \tag{2.14}$$

if $1 \leqslant i < j \leqslant n$ and $x_1 < x_2$.

Expression (2.10) enables us also to get the joint d.f.

$$F_{1,2,\ldots,n:n}(x_1,x_2,\ldots,x_n) = P\{X_{1,n} \leqslant x_1, X_{2,n} \leqslant x_2,\ldots, X_{n,n} \leqslant x_n\}.$$

One has

$$F_{1,2,\ldots,n:n}(x_1,x_2,\ldots,x_n) = n! \iiint_D \prod_{k=1}^{n} f(u_k)du_1 \cdots du_n, \tag{2.15}$$

where

$$D = \{U_1,\ldots,U_n : U_1 < U_2 < \cdots < U_n; U_1 < x_1, U_2 < x_2,\ldots, U_n < x_n\}.$$

Note that (2.15) is equivalent to the expression

$$F_{1,2,\ldots,n:n}(x_1,x_2,\ldots,x_n) = n! \iiint_{\widehat{D}} du_1 \cdots du_n, \tag{2.16}$$

where integration is over

$$\widehat{D} = \{U_1,\ldots,U_n : U_1 < U_2 < \cdots < U_n; U_1 < F(x_1), U_2 < F(x_2),\ldots, U_n < F(x_n)\}.$$

Remark 2.5. It can be proved that unlike (2.15), which needs the existence of population density function f, expression (2.16), as well as (2.9), is valid for arbitrary distribution function F.

Check your solutions

Exercise 2.1 (solution). If $x \geqslant y$, then

$$P\{M(n-1) \leqslant x, \, M(n) \leqslant y\} = P\{M(n) \leqslant y\} = H(y,y,\ldots,y).$$

Otherwise,

$$P\{M(n-1) \leqslant x, \, M(n) \leqslant y\} = P\{M(n-1) \leqslant x, \, X_n \leqslant y\} = H(x,\ldots,x,y).$$

Exercise 2.2 (answer).

$$P\{Y \leqslant x\} = 1 - H(x,\infty) - H(\infty,x) + H(x,x),$$

where

$$H(x,\infty) = P\{X_1 \leqslant x, \, X_2 < \infty\} = P\{X_1 \leqslant x\}$$

and

$$H(\infty,x) = P\{X_2 \leqslant x\}.$$

Exercise 2.3 (solution). If $x \geqslant y$, then

$$P\{Y \leqslant x, \, Z \leqslant y\} = P\{Z \leqslant y\} = H(y,y).$$

If $x < y$, then

$$P\{Y \leqslant x, \, Z \leqslant y\} = P\{X_1 \leqslant x, \, X_2 \leqslant y\} + P\{X_1 \leqslant y, \, X_2 \leqslant x\} - P\{X_1 \leqslant x, \, X_2 \leqslant x\}$$

$$= H(x,y) + H(y,x) - H(x,x).$$

Exercise 2.4 (solution). We see that

$$P\{Y \geqslant r, \, Z < s\} = P\{r \leqslant X_k < s, \, k = 1, 2, \ldots, n\}$$

$$= (P\{r \leqslant X < s\})^n = (P\{X \geqslant r\} - P\{X \geqslant s\})^n = (p^r - p^s)^n.$$

Exercise 2.5 (solution). It is easy to see that (2.6) is valid for $x = 0$. Now it suffices to prove that both sides of (2.6) have equal derivatives. The derivative of the RHS is naturally equal to

$$\frac{x^{k-1}(1-x)^{n-k}}{B(k,n-k+1)} = \frac{n!x^{k-1}(1-x)^{n-k}}{(k-1)!(n-k)!},$$

because $B(a,b) = \Gamma(a)\Gamma(b)/\Gamma(a+b)$ and the gamma function satisfies equality $\Gamma(k) = (k-1)!$ for $k = 1, 2, \ldots$.

It turns out (after some simple calculations) that the derivative of the LHS also equals

$$\frac{n!x^{k-1}(1-x)^{n-k}}{(k-1)!(n-k)!}.$$

Exercise 2.6 (solution). Let $r < s$. Denote

$$F_{r,s:n}(x_1, x_2) = P\{X_{r,n} \leqslant x_1, X_{s,n} \leqslant x_2\}.$$

If $x_2 \leqslant x_1$, then evidently

$$P\{X_{r,n} \leqslant x_1, X_{s,n} \leqslant x_2\} = P\{X_{s,n} \leqslant x_2\}$$

and

$$F_{r,s:n}(x_1, x_2) = \sum_{m=s}^{n} \binom{n}{m} (F(x_2))^m (1 - F(x_2))^{n-m} = \mathbf{1}_{F(x_2)}(s, n - s + 1).$$

Consider now the case $x_2 > x_1$. To find $F_{r,s:n}(x_1, x_2)$ let us mention that any X from the sample $X_1, X_2, \ldots, X_n$ (independently on other X's) with probabilities $F(x_1)$, $F(x_2) - F(x_1)$ and $1 - F(x_2)$ can fall into intervals $(-\infty, x_1]$, $(x_1, x_2]$, (x_2, ∞) respectively. One sees that the event $A = \{X_{r,n} \leqslant x_1, X_{s,n} \leqslant x_2\}$ is a union of some disjoint events

$$a_{i,j,n-i-j} = \{ i \text{ elements of the sample fall into } (-\infty, x_1],$$
$$j \text{ elements fall into interval } (x_1, x_2], \text{ and}$$
$$(n - i - j) \text{ elements lie to the right of } x_2 \}.$$

Recalling the polynomial distribution we obtain that

$$P\{A_{i,j,n-i-j}\} = \frac{n!}{i! j! (n - i - j)!} (F(x_1))^i (F(x_2) - F(x_1))^j (1 - F(x_2))^{n-i-j}.$$

To construct A one has to take all $a_{i,j,n-i-j}$ such that $r \leqslant i \leqslant n$, $j \geqslant 0$ and $s \leqslant i + j \leqslant n$. Hence,

$$F_{r,s:n}(x_1, x_2) = P\{A\} = \sum_{i=r}^{n} \sum_{j=\max\{0, s-i\}}^{n-i} P\{A_{i,j,n-i-j}\}$$

$$- \sum_{i=r}^{n} \sum_{j=\max\{0, s-i\}}^{n-i} \frac{n!}{i! j! (n - i - j)!} (F(x_1))^i (F(x_2) - F(x_1))^j (1 - F(x_2))^{n-i-j}.$$

Exercise 2.7 (answer). Denote for convenience, $k(0) = 0$, $k(r+1) = n + 1$, $x_0 = -\infty$ and $x_{r+1} = \infty$. Then

$$f_{k(1),k(2),\ldots,k(r):n}(x_1, x_2, \ldots, x_r) =$$

$$\begin{cases} \dfrac{n!}{\prod\limits_{m=1}^{r+1} (k(m) - k(m-1) - 1)!} \prod\limits_{m=1}^{r+1} (F(x_m) - F(x_{m-1}))^{k(m)-k(m-1)-1} \prod\limits_{m=1}^{r} f(x_m), \\ \qquad\qquad\qquad\qquad\qquad\qquad \text{if } x_1 < x_2 < \cdots < x_r, \text{ and} \\ \qquad\qquad 0, \qquad\qquad\qquad\qquad \text{otherwise.} \end{cases}$$

In particular, if $r = 2$, $1 \leqslant i < j \leqslant n$, and $x_1 < x_2$, then

$$f_{i,j:n}(x_1, x_2) = \frac{n!}{(i-1)!(j-i-1)!(n-j)!}$$

$$\times (F(x_1))^{i-1} (F(x_2) - F(x_1))^{j-i-1} (1 - F(x_2))^{n-j} f(x_1) f(x_2).$$

Chapter 3

Sample Quantiles and Ranges

We define and discuss such important statistics as sample quantiles $X_{[pn]+1,n}$, ranges $X_{n,n} - X_{1,n}$, quasi-ranges $X_{n-r+1,n} - X_{r,n}$, $r = 2, 3, \ldots, [n/2]$, midranges $(X_{n,n} + X_{1,n})/2$ and quasi-midranges $(X_{n-r+1,n} + X_{r,n})/2$. Distributions and some properties of this statistics related to order statistics are obtained.

It turns out very often that in statistical inference estimates of some unknown parameters, which are the best in some sense (efficient, robust) or satisfies useful properties (sufficient, simple and convenient for applications), have the form of order statistics or can be expressed as functions of order statistics.

Example 3.1. Let us have a sample $X_1, X_2, \ldots, X_n$ of size n from a population d.f. $F(x, \theta)$, θ being an unknown parameter, which we need to estimate using some statistic $T = T(X_1, \ldots, X_n)$. Statistic T is sufficient if it contains as much information about θ as all sample $X_1, \ldots, X_n$. Rigorously saying, T is a sufficient for θ if the conditional distribution of the vector $(X_1, \ldots, X_n)$ given $T = t$ does not depend on θ. There are some useful criteria to determine sufficient statistics. For instance, consider the case, when our population has a probability density function $f(x, \theta)$. Then T is sufficient for θ if the equality

$$f(x_1, \theta)f(x_2, \theta) \cdots f(x_n, \theta) = h(x_1, x_2, \ldots, x_n)g(\theta, T(x_1, x_2, \ldots, x_n)) \tag{3.1}$$

holds for some nonnegative functions h (which does not depend on θ) and g (which depends on θ and $T(x_1, x_2, \ldots, x_n)$ only).

Let now X have the uniform distribution on $[0, \theta]$, $\theta > 0$ being an unknown parameter. In this case $f(x, \theta) = 1/\theta$, if $0 \leqslant x \leqslant \theta$, and $f(x, \theta) = 0$, otherwise. Thus,

$$f(x_1, \theta)f(x_2, \theta) \cdots f(x_n, \theta) = \begin{cases} \dfrac{1}{\theta^n}, & 0 \leqslant x_1, x_2, \ldots, x_n \leqslant \theta, \\ 0, & \text{otherwise.} \end{cases} \tag{3.2}$$

The RHS of (3.2) can be expressed as

$$h(x_1, x_2, \ldots, x_n)g(\theta, T(x_1, x_2, \ldots, x_n)),$$

M. Ahsanullah et al., *An Introduction to Order Statistics*,
Atlantis Studies in Probability and Statistics 3,
DOI: 10.2991/978-94-91216-83-1_3, © Atlantis Press 2013

where

$$h(x_1, x_2, \ldots, x_n) = \begin{cases} 1, & \text{if } x_k \geqslant 0,\, k = 1, 2, \ldots, n, \text{ and} \\ 0, & \text{otherwise;} \end{cases}$$

$$T(x_1, x_2, \ldots, x_n) = \max\{x_1, x_2, \ldots, x_n\}$$

and

$$g(\theta, T(x_1, x_2, \ldots, x_n)) = \frac{1_{\{T(x_1, x_2, \ldots, x_n) \leqslant \theta\}}}{\theta^n}.$$

One sees that in this case the sufficient statistic has the form

$$T = \max\{X_1, \ldots, X_n\} = X_{n,n}.$$

Example 3.2. Let us again consider a sample $X_1, \ldots, X_n$ from a population, having a probability density function $f(x, \theta)$, where θ is an unknown parameter. In this situation the likelihood function is defined as

$$L(x_1, x_2, \ldots, x_n, \theta) = f(x_1, \theta) f(x_2, \theta) \cdots f(x_n, \theta). \tag{3.3}$$

To construct the maximum likelihood estimate of θ one must find such

$$\theta^* = \theta^*(x_1, \ldots, x_n),$$

which maximizes the RHS of (3.3), and take $\theta^*(x_1, \ldots, x_n)$.

Consider the case, when a sample is taken from the Laplace distribution with p.d.f.

$$f(x, \theta) = \frac{\exp(-|x - \theta|)}{2}.$$

What is the maximum likelihood estimate of θ? We see that the likelihood function has the form

$$L(x_1, x_2, \ldots, x_n, \theta) = \frac{\exp\left\{ -\sum_{k=1}^{n} |x_k - \theta| \right\}}{2^n}. \tag{3.4}$$

To maximize $L(x_1, x_2, \ldots, x_n, \theta)$ it suffices to minimize

$$\sigma(\theta) = \sum_{k=1}^{n} |x_k - \theta|.$$

Let us arrange observations $x_1, \ldots, x_n$ in the non-decreasing order: $x_{1,n} \leqslant \cdots \leqslant x_{n,n}$. Here we must distinguish two situations. At first, we consider odd values of n. Let $n = 2k + 1$, $k = 0, 1, 2, \ldots$. It is not difficult to show that $\sigma(\theta)$ decreases with respect to θ in the interval $(-\infty, x_{k+1,2k+1})$ and $\sigma(\theta)$ increases for $\theta \in (x_{k+1,2k+1}, \infty)$. It implies that $\sigma(\theta)$ attains its minimal value if θ coincides with $X_{k+1,2k+1}$. Thus, in this case order statistic $X_{k+1,2k+1}$ is the maximum likelihood estimate of the location parameter θ. If $n = 2k$,

$k = 1, 2, \ldots$, is even, then $\sigma(\theta)$ decreases in $(-\infty, x_{k,2k})$, increases in $(x_{k+1,2k}, \infty)$ and has a constant value in the interval $(x_{k,2k}, x_{k+1,2k})$. Hence $\sigma(\theta)$ attains minimal values for any $\theta \in [x_{k,2k}, x_{k+1,2k}]$. Any point of this interval can be presented as

$$\alpha x_{k,2k} + (1 - \alpha) x_{k+1,2k},$$

where $0 \leqslant \alpha \leqslant 1$. It means that any statistic of the form

$$\alpha X_{k,2k} + (1 - \alpha) X_{k+1,2k}, \quad 0 \leqslant \alpha \leqslant 1,$$

is the maximum likelihood estimate of θ. If $\alpha = 1/2$, we get the statistics

$$\frac{X_{k,2k} + X_{k+1,2k}}{2}.$$

Example 3.3. Consider a sample from the normal $N(\theta, \sigma^2)$ population. Let

$$\overline{X} = \frac{X_1 + \cdots + X_n}{n}$$

denote the sample mean. It is known that for normal distributions the vector

$$(X_1 - \overline{X}, X_2 - \overline{X}, \ldots, X_n - \overline{X})$$

and $\overline{X}$ are independent. Then the vector

$$(X_{1,n} - \overline{X}, X_{2,n} - \overline{X}, \ldots, X_{n,n} - \overline{X})$$

and $\overline{X}$ are also independent. In statistical inference, based on the normal samples, statisticians very often need to use independent estimates of location (θ) and scale (σ) parameters. The best in any respects for their purposes are independent estimates $\overline{X}$ and

$$S = \left(\frac{1}{n-1} \sum_{k=1}^{n} (X_k - \overline{X})^2 \right)^{1/2}.$$

For the sake of simplicity we can change S by another estimate of σ. Convenient analogues of S are presented by statistics

$$d(r, k, n)(X_{r,n} - X_{k,n}),$$

where $d(r, k, n)$ are correspondingly chosen normalizing constants, which provide unbiased estimation of parameter σ. Since differences $(X_{r,n} - \overline{X})$ and $(X_{s,n} - \overline{X})$ do not depend on $\overline{X}$, the statistic

$$(X_{r,n} - X_{k,n}) = (X_{r,n} - \overline{X}) - (X_{s,n} - \overline{X})$$

does not depend on $\overline{X}$ also. Thus, we see that statistics of the form $X_{r,n} - X_{k,n}$, $1 \leqslant k < r \leqslant n$, have good properties and can be used in the estimation theory.

The suggested examples show that order statistics naturally arise in statistical inference. The most popular are extreme order statistics, such as $X_{1,n}$, $X_{n,n}$, and the so-called sample quantiles. To determine sample quantiles we must recall the definition of the quantiles for random variables (or quantiles of d.f. F).

Definition 3.1. A value x_p is called a quantile of order p, $0 < p < 1$, if

$$P\{X < x_p\} \leqslant p \leqslant P\{X \leqslant x_p\}. \tag{3.5}$$

If F is the d.f. of X, then (3.5) is equivalent to the relation

$$F(x_p - 0) \leqslant p \leqslant F(x_p). \tag{3.6}$$

For continuous F, x_p is any solution of the equation

$$F(x_p) = p. \tag{3.7}$$

Note that (3.7) has a unique solution, if F is strictly increasing. Otherwise, any point of the interval $[\delta, \gamma]$, where

$$\delta = \inf\{x : F(x) = p\}$$

and

$$\gamma = \sup\{x : F(x) = p\},$$

satisfies (3.7) and may be called a quantile of order p. Very often one uses the sample d.f.

$$F_n^*(x) = \frac{1}{n} \sum_{k=1}^{n} 1_{\{X_k \leqslant x\}}$$

to estimate the population d.f. F. It is natural to take quantiles of $F_n^*(x)$ as estimates of quantiles of F. Substituting $F_n^*(x)$ instead of F to (3.6), we get the relation

$$F_n^*(x - 0) \leqslant p \leqslant F_n^*(x). \tag{3.8}$$

Let us recall now (see (1.16)) that $F_n^*(x) = k/n$, if $X_{k,n} \leqslant x < X_{k+1,n}$, $1 \leqslant k \leqslant n - 1$.
Comparing (1.16) and (3.8) one obtains that the only solution of (3.8) is $X_{k,n}$ if $(k-1)/n < p < k/n$, $k = 1, 2, \ldots, n$. If $p = k/n$, $k = 1, 2, \ldots, n - 1$, then any $x \in [X_{k,n}, X_{k+1,n}]$ satisfies (3.8). Hence, if np is an integer, then any statistic of the form

$$\alpha X_{pn,n} + (1 - \alpha)X_{pn+1,n}, \quad 0 \leqslant \alpha \leqslant 1,$$

including $X_{pn,n}$, $X_{pn+1,n}$, $(X_{pn,n} + X_{pn+1,n})/2$ as possible options, can be regarded as a sample quantile. Otherwise, (3.8) has the unique solution $X_{[pn]+1,n}$. Thus, the evident simple definition of the sample quantile of order p, $0 < p < 1$, which covers both situations, is given as $X_{[pn]+1,n}$.

Example 3.4. We return to the case, when np is an integer. Indeed, if a size of the sample is large, there is no essential difference between all possible versions of sample quantiles and we can take the simplest of them, say $X_{pn+1,n}$. For small sizes of samples different definitions can give distinguishable results of statistical procedures. Hence in any concrete situation we must choose the best (in some sense) of the statistics

$$\alpha X_{pn,n} + (1 - \alpha)X_{pn+1,n}, \quad 0 \leqslant \alpha \leqslant 1.$$

One of the possible criterions of the optimal choice is the unbiasedness of the statistic. Consider the case, when the sample quantile is used to estimate the quantile of order p for the uniform on the interval $[0, 1]$ distribution. The statistic

$$\alpha X_{pn,n} + (1 - \alpha)X_{pn+1,n}$$

is unbiased in this case, if the following equality holds:

$$E(\alpha X_{pn,n} + (1 - \alpha)X_{pn+1,n}) = p. \tag{3.9}$$

In Chapter 8 we will find that

$$EX_{k,n} = \frac{k}{n+1}, \quad 1 \leqslant k \leqslant n, \tag{3.10}$$

for the uniform $U([0, 1])$ distribution. From (3.9) and (3.10) we obtain that $\alpha = 1 - p$.

The most important of sample quantiles is the sample median, which corresponds to the case $p = 1/2$. If $n = 2k + 1$, $k = 1, 2, \ldots$, then the sample median is defined as $X_{k+1,2k+1}$. For even ($n = 2k$, $k = 1, 2, \ldots$) size of a sample any statistics of the form

$$\alpha X_{k,2k} + (1 - \alpha)X_{k+1,2k}, \quad 0 \leqslant \alpha \leqslant 1,$$

may be regarded as the sample median. Sample medians are especially good for estimation of the location parameter in the case, when the population distribution is symmetric.

We say that X is symmetric random variable if X and $-X$ have the same distribution. Analogously, X is symmetric with respect to some location parameter θ, if $X - \theta$ and $\theta - X$ have the same distribution. If X is symmetric with respect to some value θ, then $X - \theta$ is simply symmetric. If X is symmetric with respect to θ and there exists the expectation of X, then EX equals θ, as well as the median of X. Hence, if θ is an unknown parameter, we can use different estimates for θ, the sample mean and the sample median among them. Moreover, there are situations (see, for instance, exercise 3.2), when the sample median is the best estimate in some sense.

Exercise 3.1. Show that if X has symmetric distribution then equality

$$x_p = -x_{1-p}, \quad 0 < p < 1,$$

holds for quantiles of order p and $1 - p$.

Exercise 3.2. Let $X_{1,n}, X_{2,n}, \ldots, X_{n,n}$ be order statistics, corresponding to a continuous d.f. F, and let x_p and x_{1-p} denote quantiles of order p and $1 - p$ respectively for F. Show that equality

$$P\{X_{[\alpha n],n} \leqslant x_p\} + P\{X_{[(1-\alpha)n],n} \leqslant x_{1-p}\} = 1$$

holds for any $0 < \alpha < 1$ in the case, when αn is not an integer, while the relation

$$P\{X_{\alpha n,n} \leqslant x_p\} + P\{X_{(1-\alpha)n+1,n} \leqslant x_{1-p}\} = 1$$

is valid for the case, when αn is an integer.

The statement of the next exercise arises to van der Vaart's (1961) paper.

Exercise 3.3. Let $X_{k+1,2k+1}$ be a sample median based on a sample of odd size from a distribution with a continuous d.f. F. Show that the median of $X_{k+1,2k+1}$ coincides with the median of the population distribution.

Exercise 3.4. Let $X_{k+1,2k+1}$ be a sample median, corresponding to a sample of odd size. Show that $X_{k+1,2k+1}$ has a symmetric distribution if and only if the population distribution is symmetric.

As we mentioned above, sample medians may be good estimates of the location parameter. One more type of statistics, which are used for estimation of the location parameter, is presented by different midranges. The classical midrange is defined as

$$\frac{X_{1,n} + X_{n,n}}{2},$$

while the so-called quasi-midranges are given as

$$\frac{X_{k,n} + X_{n-k+1,n}}{2}, \quad k = 2, \ldots, \left[\frac{n}{2}\right].$$

We can see that the sample median

$$\frac{X_{k,2k} + X_{k+1,2k}}{2}$$

also presents one of quasi-midranges. As a measure of the population spread, a statistician can use ranges $X_{n,n} - X_{1,n}$ and quasi-ranges

$$X_{n-k+1,n} - X_{k,n}, \quad k = 2, \ldots, \left[\frac{n}{2}\right].$$

In example 3.3 we found that any quasi-range $X_{n-k+1,n} - X_{k,n}$ (as well as range $X_{n,n} - X_{1,n}$) and the sample mean $\overline{X} = (X_1 + \cdots + X_n)/n$ are independent for the normal distribution. To use ranges and midranges in statistical inference we need to know distributions of these statistics.

Example 3.5. We suppose that our population has p.d.f. f and try to find probability density functions of ranges

$$W_n = X_{n,n} - X_{1,n}, \quad n = 2, 3, \ldots.$$

Substituting $i = 1$ and $j = n$ to (2.14), we get the joint pdf of order statistics $X_{1,n}$ and $X_{n,n}$:

$$f_{1,n:n}(x,y) = n(n-1)(F(y) - F(x))^{n-2} f(x)f(y), \quad x < y, \tag{3.11}$$

and

$$f_{1,n:n}(x,y) = 0, \quad \text{if } x \geqslant y.$$

Consider the linear change of variables $(u,v) = (x, y - x)$ with the unit Jacobian, which corresponds to the passage to random variables $U = X_{1,n}$ and $V = X_{n,n} - X_{1,n} > 0$. Now (3.11) implies that random variables U and V have the joint density function

$$f_{U,V}(u,v) = \begin{cases} n(n-1)(F(u+v) - F(u))^{n-2} f(u) f(u+v), & -\infty < u < \infty, \quad v > 0, \\ 0, & \text{otherwise.} \end{cases} \tag{3.12}$$

Integrating (3.12) with respect to u, we obtain that the range $X_{n,n} - X_{1,n}$ has the density

$$f_V(v) = n(n-1) \int_{-\infty}^{\infty} (F(u+v) - F(u))^{n-2} f(u) f(u+v) du, \quad v > 0. \tag{3.13}$$

One more integration (now with respect to v) enables us to get the distribution function of the range:

$$F_V(x) = P\{X_{n,n} - X_{1,n} \leqslant x\} = \int_0^x F_V(v) dv$$

$$= n \int_{-\infty}^{\infty} f(u) \left(\int_0^x d((F(u+v) - F(u))^{n-1}) \right) du$$

$$= n \int_{-\infty}^{\infty} (F(u+x) - F(u))^{n-1} f(u) du, \quad x > 0. \tag{3.14}$$

Exercise 3.5. Find the distribution of quasi-range

$$X_{n-r+1,n} - X_{r,n}, \quad r = 1, 2, \ldots, [n/2],$$

when

$$f(x) = 1, \quad a \leqslant x \leqslant a + 1.$$

Exercise 3.6. Let

$$F(x) = \max\{0, 1 - \exp(-x)\}.$$

Show that $U = X_{1,n}$ and the range $V = X_{n,n} - X_{1,n}$ are independent.

Exercise 3.7. Let $f(x)$ be a population density function and

$$V = \frac{X_{r,n} + X_{n-r+1,n}}{2}$$

be the corresponding quasi-midrange. Find the p.d.f. of V.

Example 3.6. We will find now the joint distribution of $W = X_{n,n} - X_{1,n}$ and

$$V = \frac{X_{1,n} + X_{n,n}}{2}.$$

Consider (3.11) and make the linear change of variables

$$(w, v) = \left(y - x, \frac{x+y}{2} \right)$$

with the unit Jacobian, which corresponds to the passage to random variables $W > 0$ and V. After noting that

$$x = v - \frac{w}{2} \quad \text{and} \quad y = v + \frac{w}{2},$$

we get the joint probability density function of W and V:

$$f_{W,V}(w,v) = n(n-1) \left(F\left(v+\frac{w}{2}\right) - F\left(v-\frac{w}{2}\right) \right)^{n-2} f\left(v-\frac{w}{2}\right) f\left(v+\frac{w}{2}\right),$$

$$-\infty < v < \infty, \quad w > 0. \quad (3.15)$$

Remark 3.1. Consider the joint distribution of range W and midrange V in the case, when $n = 2$ and

$$f(x) = \frac{1}{\sqrt{2\pi}} \exp\left(-\frac{x^2}{2} \right).$$

From (3.15) we have

$$f_{W,V}(w,v) = \frac{1}{\pi} \exp(-v^2) \exp\left(-\frac{w^2}{4} \right), \quad -\infty < v < \infty, \quad w > 0. \quad (3.16)$$

Equality (3.16) means that W and V are independent. This fact is not surprising so far as for $n = 2$ the midrange coincides with the sample mean

$$\overline{X} = \frac{X_1 + X_2}{2}$$

and we know from example 3.3 that the ranges and sample means

$$\overline{X} = \frac{X_1 + X_2 + \cdots + X_n}{n}$$

are independent for normal distributions.

Check your solutions

Exercise 3.1 (solution). If X is symmetrically distributed, then

$$P\{X < x\} = P\{X > -x\}$$

and

$$P\{X \leqslant x\} = P\{X \geqslant -x\}.$$

Let x_p be a quantile of order p. It means (see (3.6)) that

$$P\{X < x_p\} \leqslant p \leqslant P\{X \leqslant x_p\}.$$

It follows now from the relations given above that

$$P\{X > -x_p\} \leqslant p \leqslant P\{X \geqslant -x_p\}.$$

The latter inequalities can be rewritten in the form

$$P\{X < -x_p\} \leqslant 1 - p \leqslant P\{X \leqslant -x_p\},$$

which confirms that $-X_p$ satisfies the definition of the quantile of order $1 - p$. Note that if any point of some interval $[a,b]$ is a quantile of order p for symmetric distribution, then any point of interval $[-b,-a]$ is a quantile of order $1 - p$.

Exercise 3.2 (solution). Recalling that $F(x_p) = p$ for continuous d.f., we have from (2.5) that

$$P\{X_{[\alpha n],n} \leqslant x_p\} = \sum_{m=[\alpha n]}^{n} \binom{n}{m} (F(x_p))^m (1 - F(x_p))^{n-m}$$

$$= \sum_{m=[\alpha n]}^{n} \binom{n}{m} p^m (1 - p)^{n-m}$$

and similarly

$$P\{X_{[(1-\alpha)n],n} \leqslant x_{1-p}\} = \sum_{m=[(1-\alpha)n]}^{n} \binom{n}{m} (1 - p)^m p^{n-m}$$

$$= \sum_{m=0}^{n-[(1-\alpha)n]} \binom{n}{n - m} (1 - p)^{n-m} p^m$$

$$= \sum_{m=0}^{n-[(1-\alpha)n]} \binom{n}{m} p^m (1 - p)^{n-m},$$

so far as

$$\binom{n}{n - m} = \binom{n}{m}.$$

Consider values $r = [\alpha n]$ and $s = [(1 - \alpha)n]$. It is easy to see that $r + s = n$, if αn is an integer, and $r + s = n - 1$ in the opposite case. Thus, if αn is not an integer, we obtain that

$$n - [(1 - \alpha)n] = r - 1 = [\alpha n] - 1$$

and

$$P\{X_{[\alpha n],n} \leqslant x_p\} + P\{X_{[(1-\alpha)n],n} \leqslant x_{1-p}\} = \sum_{m=0}^{n} \binom{n}{m} p^m (1-p)^{n-m} = (p + (1-p))^n = 1.$$

If αn is an integer, then $n - [(1 - \alpha)n] = r = [\alpha n]$ and we get one more term in our sum. In this case

$$P\{X_{\alpha n,n} \leqslant x_p\} + P\{X_{(1-\alpha)n,n} \leqslant x_{1-p}\} = 1 + \binom{n}{\alpha n} p^{\alpha n} (1-p)^{(1-\alpha)n} > 1.$$

Some understandable change gives us the following equality:

$$P\{X_{\alpha n,n} \leqslant x_p\} + P\{X_{(1-\alpha)n+1,n} \leqslant x_{1-p}\} = 1.$$

Exercise 3.3 (solution). Let μ be a population median. It means that

$$F(\mu) = \frac{1}{2}.$$

It follows from (2.5) that

$$P\{X_{k+1,2k+1} \leqslant \mu\} = \sum_{m=k+1}^{2k+1} \binom{2k+1}{m} (F(\mu))^m (1 - F(\mu))^{2k+1-m}$$

$$= \sum_{m=k+1}^{2k+1} \binom{2k+1}{m} \left(\frac{1}{2}\right)^m \left(\frac{1}{2}\right)^{2k+1-m} = \left(\frac{1}{2}\right)^{2k+1} \sum_{m=k+1}^{2k+1} \binom{2k+1}{m} = \frac{1}{2},$$

so far as

$$\sum_{m=k+1}^{2k+1} \binom{2k+1}{m} = \sum_{m=0}^{k} \binom{2k+1}{m}$$

and hence,

$$\sum_{m=k+1}^{2k+1} \binom{2k+1}{m} = \frac{1}{2} \sum_{m=0}^{2k+1} \binom{2k-1}{m} (2k+1) = \frac{1}{2} (1+1)^{2k+1} = 2^{2k}.$$

Exercise 3.4 (solution). We can use equality (2.8), which in our situation has the form

$$F_{k+1,2k+1}(x) = I_{F(x)}(k+1, k+1) = \frac{1}{B(k+1, k+1)} \int_0^{F(x)} t^k (1-t)^k dt,$$

where

$$B(k+1, k+1) = \int_0^1 t^k (1-t)^k dt.$$

If the population distribution is symmetric,

$$F(-x) = 1 - F(x-0)$$

and then for any X we have

$$1 - F_{k+1,2k+1}(x-0) = 1 - \frac{1}{B(k+1,k+1)} \int_0^{F(x-0)} t^k(1-t)^k dt$$

$$= \frac{1}{B(k+1,k+1)} \int_{F(x-0)}^1 t^k(1-t)^k dt$$

$$= \frac{1}{B(k+1,k+1)} \int_{1-F(-x)}^1 t^k(1-t)^k dt$$

$$= \frac{1}{B(k+1,k+1)} \int_0^{F(-x)} t^k(1-t)^k dt = F_{k+1,2k+1}(-x)$$

and this means that $X_{k+1,2k+1}$ has a symmetric distribution.

Let now admit that $X_{k+1,2k+1}$ is a symmetric random variable. Then, for any x,

$$1 - F_{k+1,2k+1}(x-0) = F_{k+1,2k+1}(-x)$$

and this is equivalent to the relation

$$I_{F(-x)}(k+1,k+1) + I_{F(x-0)}(k+1,k+1) = 1.$$

It is not difficult to see that the latter equality, in its turn, is equivalent to the relation

$$\int_0^{F(-x)} t^k(1-t)^k dt + \int_{1-F(x-0)}^1 t^k(1-t)^k dt = B(k+1,k+1) = \int_0^1 t^k(1-t)^k dt,$$

which immediately implies that

$$F(-x) = 1 - F(x-0).$$

Hence, the population distribution is symmetric.

Exercise 3.5 (hint and answer). Consider (2.14) with

$$i = r, \ j = n-r+1, \ f(x) = 1, \ a \leqslant x \leqslant a+1, \ F(x) = x-a, \ a \leqslant x \leqslant a+1$$

and use the linear change of variables $(u,v) = (x, y-x)$. It will give you the joint density function of $X_{r,n}$ and $W(r) = X_{n-r+1,n} - X_{r,n}$. Now the integration with respect to u enables you to get the density function of $W(r)$.

It turns out, that the density function of $W(r)$ does not depend on a and has the form

$$f_{W(r)}(x) = \frac{x^{n-2r}(1-x)^{2r-1}}{B(n-2r+1,2r)}, \quad 0 \leqslant x \leqslant 1,$$

i.e. $W(r)$ has the beta distribution with parameters $n-2r+1$ and $2r$.

Exercise 3.6 (solution). In this case, recalling (3.12), we have that for any $u > 0$ and $v > 0$, the joint density function is given as follows:

$$f_{U,V}(u,v) = n(n-1)(\exp(-u) - \exp(-(u+v)))^{n-2} \exp(-u) \exp(-u-v)$$
$$= n(n-1)\exp(-nu)(1 - \exp(-v))^{n-2} \exp(-v)$$
$$= h(u)g(v),$$

where

$$h(u) = n \exp(-nu)$$

and

$$g(v) = (n-1)(1 - \exp(-v))^{n-2} \exp(-v).$$

The existence of the factorization

$$f_{U,V}(u,v) = h(u)g(v)$$

suffices to state that U and V are independent random variables. Indeed, one can check that in fact $h(u)$ and $g(v)$ are densities of U and V respectively.

Exercise 3.7 (solution). Substituting $i = r$ and $j = n-r+1$ to (2.14), we get the joint pdf of order statistics $X_{r,n}$ and $X_{n-r+1,n}$:

$$f_{r,n-r+1:n}(x,y) = \begin{cases} c(r,n)(F(x))^{r-1}(F(y) - F(x))^{n-2r}(1 - F(y))^{r-1}f(x)f(y), & x < y, \\ 0, & \text{otherwise,} \end{cases}$$

where

$$c(r,n) = n!/(r-1)!(n-2r)!(r-1)!.$$

Consider the linear change of variables $(u,v) = (x,(x+y)/2)$, which corresponds to the passage to random variables $U = X_{1,n}$ and $V = (X_{r,n} + X_{n-r+1,n})/2$.

After noting that the Jacobian of this transformation is $1/2$, $x = u$, $y = 2v - u$ and the inequality $y > x$ means that $v > u$, we obtain that the joint p.d.f. of U and V is given as follows:

$$f_{U,V}(u,v) = \begin{cases} 2c(r,n)(F(u))^{r-1}(F(2v - u) - F(u))^{n-2r}(1 - F(2v - u))^{r-1}f(u)f(2v - u), \\ \qquad\qquad\qquad\qquad\qquad \text{if } u < v, \\ 0, \qquad\qquad\qquad\qquad \text{if } u \geqslant v. \end{cases}$$

Integration with respect to u enables us to get the density function of V:

$$f_V(v) = 2c(r,n) \int_{-\infty}^{v} (F(u))^{r-1}(F(2v - u) - F(u))^{n-2r}(1 - F(2v - u))^{r-1}f(u)f(2v - u)du,$$

$$-\infty < v < \infty.$$

In particular, the p.d.f. of midrange

$$M = \frac{X_{1,n} + X_{n,n}}{2}$$

has the form

$$f_m(v) = 2n(n-1) \int_{-\infty}^{v} (F(2v-u) - F(u))^{n-2} f(u) f(2v-u) du, \quad -\infty < v < \infty.$$

while the p.d.f.'s of the quasi-midranges

$$\frac{X_{k,2k} + X_{k+1,2k}}{2}, \quad k = 1, 2, \ldots,$$

which coincide with sample medians for samples of even size, are given as follows:

$$f_V(v) = \frac{2(2k)!}{((k-1)!)^2} \int_{-\infty}^{v} (F(u))^{k-1} (1 - F(2v-u))^{k-1} f(u) f(2v-u) du, \quad -\infty < v < \infty.$$

Chapter 4

Representations for Order Statistics

We prove some important relations, which enable us to express order statistics, in terms of sums or products of independent random variables. Let $U_{1,n} \leqslant \cdots \leqslant U_{n,n}$ be the uniform order statistics (corresponding to the d.f. $F(x) = x$, $0 \leqslant x \leqslant 1$) and $Z_{1,n} \leqslant \cdots \leqslant Z_{n,n}$ be the exponential order statistics related to d.f. $F(x) = 1 - \exp(-x)$, $x > 0$. It turns out that, in particular, the following equalities are valid:

$$U_{k,n} \stackrel{d}{=} W_k^{1/k} W_{k+1}^{1/(k+1)} \cdots W_n^{1/n} \stackrel{d}{=} \frac{v_1 + \cdots + v_k}{v_1 + \cdots + v_{n+1}}$$

and

$$Z_{k,n} \stackrel{d}{=} \frac{v_1}{n} + \frac{v_2}{n-1} + \cdots + \frac{v_k}{n-k+1}, \quad k = 1, 2, \ldots, n,$$

where $W_1, W_2, \ldots$ and $v_1, v_2, \ldots$ are two sequences of independent identically distributed random variables, having the uniform on $[0, 1]$ distribution (in the first case) and the standard exponential $E(1)$ distribution (in the second case). In the general case, when F is any d.f., we introduce the inverse function

$$G(s) = \inf\{x : F(x) \geqslant s\}, \quad 0 < s < 1,$$

and express order statistics $X_{k,n}$ via the uniform or exponential order statistics as follows:

$$X_{k,n} \stackrel{d}{=} G(U_{k,n}) \stackrel{d}{=} G(1 - \exp(-Z_{k,n})).$$

Due to these relations we have the following useful representations for order statistics in the general case:

$$X_{k,n} \stackrel{d}{=} G\left(\frac{v_1 + \cdots + v_k}{v_1 + \cdots + v_{n+1}}\right)$$

$$\stackrel{d}{=} G\left(1 - \exp\left(-\left(\frac{v_1}{n} + \frac{v_2}{n-1} + \cdots + \frac{v_k}{n-k+1}\right)\right)\right), \quad k = 1, 2, \ldots, n.$$

M. Ahsanullah et al., *An Introduction to Order Statistics*,
Atlantis Studies in Probability and Statistics 3,
DOI: 10.2991/978-94-91216-83-1_4, © Atlantis Press 2013

Random variables $X_1, X_2, \ldots, X_n$ lose their original independence property being arranged in nondecreasing order. It is evident that order statistics tied by inequalities $X_{1,n} \leqslant X_{2,n} \leqslant \cdots \leqslant X_{n,n}$ are dependent. All situations, when we can express order statistics as functions of sums of independent terms, are very important for statisticians. In the sequel we will use the special notation $U_{1,n} \leqslant \cdots \leqslant U_{n,n}$ for the uniform order statistics (corresponding to the d.f. $F(x) = x$, $0 \leqslant x \leqslant 1$) and the notation $Z_{1,n} \leqslant \cdots \leqslant Z_{n,n}$ for the exponential order statistics ($F(x) = 1 - \exp(-x)$, $x \geqslant 0$). We will prove some useful representations for exponential and uniform order statistics. At first we will show how different results for $U_{k,n}$ and $Z_{k,n}$ can be rewritten for order statistics from arbitrary distribution. For any d.f. F we determine the inverse function

$$G(s) = \inf\{x : F(x) \geqslant s\}, \quad 0 < s < 1. \tag{4.1}$$

Exercise 4.1. Let $F(x)$ be a continuous d.f. of a random variable X. Show that in this case

$$F(G(x)) = x, \quad 0 < x < 1,$$

and $Y = F(X)$ has the uniform distribution on interval $[0, 1]$.

Remark 4.1. In exercise 4.1 we proved, in particular, that the inverse function $G(s)$ provides the equality

$$F(G(s)) = s, \quad 0 < s < 1,$$

if F is a continuous d.f. Moreover, it is not difficult to see that the dual equality

$$G(F(s)) = s, \quad -\infty < s < \infty,$$

holds for any s, where $F(s)$ strongly increases.

Remark 4.2. The second statement of exercise 4.1 is very important. In fact, for any random variable with a continuous d.f. F we have equality

$$F(X) \overset{d}{=} U, \tag{4.2}$$

where

$$Y \overset{d}{=} Z$$

denotes that random variables (or random vectors) Y and Z have the same distribution, U in (4.2) being a random variable, which has the uniform on $[0, 1]$ distribution. Indeed, (4.2) fails if F has jump points, since then the values of $F(X)$, unlike U, do not cover all interval $[0, 1]$.

Exercise 4.2. Let X take on values $1, 2$ and 3 with equal probabilities $1/3$. Find $G(x)$, $0 < x < 1$, and the distribution of $G(U)$, where U has the uniform distribution on $[0, 1]$.

Example 4.1. Now consider a more general case than in exercise 4.2. Take an arbitrary random variable X with a d.f. F. Let G be the inverse of F. It is not difficult to see that inequality

$$G(s) \leqslant z, \quad 0 < s < 1,$$

is equivalent to the inequality $s \leqslant F(z)$. Hence, the events $\{G(U) \leqslant z\}$ and $\{U \leqslant F(z)\}$ have the same probability. Thus,

$$P\{G(U) \leqslant z\} = P\{U \leqslant F(z)\} = F(z). \tag{4.3}$$

Remark 4.3. It follows from (4.3) that relation

$$X \overset{d}{=} G(U), \tag{4.4}$$

where G is the inverse of d.f. F, holds for any random variable, while the dual equality

$$F(X) \overset{d}{=} U$$

is valid for random variables with continuous distribution functions.

Let us take a sample $X_1, X_2, \ldots, X_n$ and order statistics $X_{1,n} \leqslant \cdots \leqslant X_{n,n}$ corresponding to a d.f. F and consider random variables

$$Y_k = F(X_k), \quad k = 1, 2, \ldots, n.$$

Let now $Y_{1,n}, \ldots, Y_{n,n}$ be order statistics based on $Y_1, \ldots, Y_n$. Since F is a monotone function, it does not disturb the ordering of X's and hence the vector $(Y_{1,n}, \ldots, Y_{n,n})$ coincides with the vector $(F(X_{1,n}), \ldots, F(X_{n,n}))$. If F is a continuous d.f., then, as we know from (4.2), independent random variables $Y_1, \ldots, Y_n$ are uniformly distributed on interval $[0, 1]$ and hence the vector $(Y_{1,n}, \ldots, Y_{n,n})$ has the same distribution as the vector of uniform order statistics $(U_{1,n}, \ldots, U_{n,n})$. All the saying enables us to write that

$$(F(X_{1,n}), \ldots, F(X_{n,n})) \overset{d}{=} (U_{1,n}, \ldots, U_{n,n}). \tag{4.5}$$

Taking into account (4.4), we similarly have the following dual equality

$$(X_{1,n}, \ldots, X_{n,n}) \overset{d}{=} (G(U_{1,n}), \ldots, G(U_{n,n})), \tag{4.6}$$

which is valid (unlike (4.5)) for any distribution.

Example 4.2. Let now $X_{1,n} \leqslant \cdots \leqslant X_{n,n}$ and $Y_{1,n} \leqslant \cdots \leqslant Y_{n,n}$ be order statistics corresponding to an arbitrary d.f. F and a continuous d.f. H correspondingly. Let also G be the inverse of F. Combining relations (4.5) for Y's and (4.6) for X's, one gets the following equality, which ties two sets of order statistics:

$$(X_{1,n}, \ldots, X_{n,n}) \overset{d}{=} (G(H(Y_{1,n})), \ldots, G(H(Y_{n,n}))). \tag{4.7}$$

For instance, if we compare arbitrary order statistics $X_{1,n}, \ldots, X_{n,n}$ and exponential order statistics $Z_{1,n}, \ldots, Z_{n,n}$, then

$$H(x) = 1 - \exp(-x), \quad x > 0,$$

and (4.7) can be rewritten as

$$(X_{1,n}, \ldots, X_{n,n}) \overset{d}{=} (G(1 - \exp(-Z_{1,n})), \ldots, G(1 - \exp(-Z_{n,n}))). \tag{4.8}$$

Remark 4.4. Indeed, analogous results are valid for any monotone increasing function $R(x)$ (no necessity to suppose that R is a distribution function). Namely, if

$$Y_k = R(X_k), \quad k = 1, 2, \ldots, n,$$

then the corresponding order statistics based on Y's and X's satisfy the relation

$$(Y_{1,n}, \ldots, Y_{n,n}) \overset{d}{=} (R(X_{1,n}), \ldots, R(X_{n,n})). \tag{4.9}$$

If R is a monotone decreasing function, then transformation $R(X)$ changes the ordering of the original X's and we have the following equality:

$$(Y_{1,n}, \ldots, Y_{n,n}) \overset{d}{=} (R(X_{n,n}), \ldots, R(X_{1,n})). \tag{4.10}$$

Now we consider some special distributions, exponential and uniform among them. At first we consider order statistics based on exponential distributions.

Exercise 4.3. Let Z_1 and Z_2 be independent and have exponential distributions with parameters λ and μ respectively. Show that random variables

$$V = \min\{Z_1, Z_2\}$$

and

$$W = \max\{Z_1, Z_2\} - \min\{Z_1, Z_2\}$$

are independent.

Example 4.3. From exercise 4.3 we see that if $Z_{1,2}$ and $Z_{2,2}$ are exponential order statistics based on a sample Z_1 and Z_2 from the standard $E(1)$ exponential distribution, then random variables

$$Z_{1,2} = \min(Z_1, Z_2)$$

and

$$Z_{2,2} - Z_{1,2} = \max\{Z_1, Z_2\} - \min\{Z_1, Z_2\}$$

are independent. It turns out that one can prove a more general result for differences of exponential order statistics.

We consider exponential order statistics $Z_{0,n}, Z_{1,n}, \ldots, Z_{n,n}$, where $Z_{0,n} = 0$ is introduced for our convenience, and differences

$$V_k = Z_{k,n} - Z_{k-1,n}, \quad k = 1, 2, \ldots, n.$$

It appears that $V_1, V_2, \ldots, V_n$ are mutually independent random variables. To prove this, we recall (see (2.10)) that the joint probability density function of order statistics $X_{1,n}, \ldots, X_{n,n}$, corresponding to a distribution with a density function f, has the form

$$f_{1,2,\ldots,n:n}(x_1, x_2, \ldots, x_n) = n! \prod_{k=1}^{n} f(x_k), \quad -\infty < x_1 < x_2 < \cdots < x_n < \infty,$$

otherwise it equals zero. In our case

$$f(x) = \exp(-x), \quad x \geqslant 0,$$

and

$$f_{1,2,\ldots,n:n}(x_1, x_2, \ldots, x_n) = n! \exp(-(x_1 + x_2 + \cdots + x_n)),$$
$$0 \leqslant x_1 < x_2 < \cdots < x_n < \infty. \tag{4.11}$$

By the linear change

$$(v_1, v_2, \ldots, v_n) = (x_1, x_2 - x_1, \ldots, x_n - x_{n-1})$$

with the unit Jacobian, taking into account that

$$(x_1 + x_2 + \cdots + x_n) = nv_1 + (n-1)v_2 + \cdots + 2v_{n-1} + v_n,$$

we obtain that the joint p.d.f. of differences $V_1, V_2, \ldots, V_n$ is of the form

$$f(v_1, v_2, \ldots, v_n) = \prod_{k=1}^{n} g_k(v_k), \quad v_1 > 0, \ v_2 > 0, \ldots, \ v_n > 0, \tag{4.12}$$

where

$$g_k(v) = (n-k+1)\exp(-(n-k+1)v), \quad v > 0,$$

is the density function of the exponential $E(1/(n-k+1))$ distribution. Evidently, (4.12) means that $V_1, V_2, \ldots, V_n$ are independent. Moreover, in fact, we obtained that the vector $(V_1, V_2, \ldots, V_n)$ has the same distribution as the vector

$$\left(\frac{v_1}{n}, \frac{v_2}{n-1}, \ldots, \frac{v_{n-1}}{2}, v_n \right),$$

where $v_1, v_2, \ldots, v_n$ are independent random variables having the standard $E(1)$ exponential distribution. Let us write this fact as

$$(V_1, V_2, \ldots, V_n) \overset{d}{=} \left(\frac{v_1}{n}, \frac{v_2}{n-1}, \ldots, v_n \right). \tag{4.13}$$

Remark 4.5. The following important equalities are evident corollaries of (4.13):

$$(nV_1, (n-1)V_2, \ldots, V_n) \overset{d}{=} (v_1, v_2, \ldots, v_n) \tag{4.14}$$

i.e., normalized differences

$$(n-k+1)V_k, \quad k = 1, 2, \ldots,$$

are independent and have the same exponential $E(1)$ distribution;

$$(Z_{1,n}, Z_{2,n}, \ldots, Z_{n,n}) \overset{d}{=} \left(\frac{v_1}{n}, \frac{v_1}{n} + \frac{v_2}{n-1}, \ldots, \frac{v_1}{n} + \frac{v_2}{n-1} + \cdots + \frac{v_{n-1}}{2} + v_n \right). \tag{4.15}$$

Exercise 4.4. For exponential order statistics $Z_{1,n}, Z_{2,n}, \ldots, Z_{n,n}$ show that if $c_1 + \cdots + c_n = 0$, then $Z_{1,n}$ and a linear combination of order statistics

$$L = c_1 Z_{1,n} + c_2 Z_{2,n} + \cdots + c_n Z_{n,n}$$

are independent.

Now we will study the structure of the uniform order statistics $U_{k,n}$, $1 \leqslant k \leqslant n$.

Example 4.4. Taking into account relations (4.5) and (4.15) we get the following equalities:

$$(U_{1,n}, \ldots, U_{n,n}) \overset{d}{=} (1 - \exp(-Z_{1,n}), \ldots, 1 - \exp(-Z_{n,n}))$$

$$\overset{d}{=} \left(1 - \exp\left(-\frac{v_1}{n} \right), \ldots, 1 - \exp\left(-\left(\frac{v_1}{n} + \frac{v_2}{n-1} + \cdots + \frac{v_{n-1}}{2} + v_n \right) \right) \right), \tag{4.16}$$

where $v_1, \ldots, v_n$ are independent and have the standard $E(1)$ exponential distribution.

Recalling (4.2) we see that

$$(\exp(-v_1), \ldots, \exp(-v_n)) \overset{d}{=} (1 - U_1, \ldots, 1 - U_n)$$

$$\overset{d}{=} (W_1, \ldots, W_n)$$

$$\overset{d}{=} (W_n, \ldots, W_1), \tag{4.17}$$

where $U_1, U_2, \ldots, U_n$ as well as $W_1, W_2, \ldots, W_n$ are independent uniformly distributed on $[0,1]$ random variables. Then (4.16) can be rewritten as

$$(1 - U_{1,n}, 1 - U_{2,n}, \ldots, 1 - U_{n,n}) \overset{d}{=} (W_n^{1/n}, W_n^{1/n} W_{n-1}^{1/(n-1)}, \ldots, W_n^{1/n} W_{n-1}^{1/(n-1)} \cdots W_2^{1/2} W_1). \tag{4.18}$$

The standard uniform distribution on $[0,1]$ is symmetric with respect to the point $1/2$. This enables us to state that

$$(1 - U_{1,n}, 1 - U_{2,n}, \ldots, 1 - U_{n,n}) \overset{d}{=} (U_{n,n}, U_{n-1,n}, \ldots, U_{1,n}). \tag{4.19}$$

More strongly one can use (4.10) with the function $R(x) = 1 - x$ and get (4.19). Combining now (4.18) and (4.19) we obtain that

$$(U_{1,n}, U_{2,n}, \ldots, U_{n,n}) \overset{d}{=}$$

$$(W_1 W_2^{1/2} \cdots W_{n-1}^{1/(n-1)} W_n^{1/n}, W_2^{1/2} \cdots W_{n-1}^{1/(n-1)} W_n^{1/n}, \ldots, W_n^{1/n}). \tag{4.20}$$

Thus, we see that any uniform order statistics $U_{k,n}$ is represented in (4.20) as the product of powers of independent uniformly distributed random variables as follows:

$$U_{k,n} \overset{d}{=} W_k^{1/k} W_{k+1}^{1/(k+1)} \cdots W_n^{1/n}, \quad k = 1, 2, \ldots, n. \tag{4.21}$$

Exercise 4.5. Show that for any $n = 2, 3, \ldots$, ratios

$$V_k = (U_{k,n}/U_{k+1,n})^k, \quad k = 1, 2, \ldots, n,$$

where $U_{n+1,n} = 1$, are independent and have the same uniform distribution on $[0,1]$.

Exercise 4.6. Let $U_{k,n}$, $1 \leqslant k \leqslant n$, $n = 1, 2, \ldots$ denote order statistics based on the sequence of independent, uniformly on the interval $[0,1]$ distributed random variables $U_1, U_2, \ldots$ and $V_{k,n}$, $1 \leqslant k \leqslant n$, be order statistics corresponding to the uniformly distributed random variables $V_1, V_2, \ldots, V_n$, where V's and U's are also independent. Show that the following equality holds for any $1 \leqslant m < n$:

$$(U_{1,n}, \ldots, U_{m,n}) \overset{d}{=} (U_{1,m} V_{m+1,n}, \ldots, U_{m,m} V_{m+1,n}). \tag{4.22}$$

Example 4.5. Taking into account (4.20) we observe that any uniform order statistic $U_{k,n}$ satisfies the following equality:

$$U_{k,n} \overset{d}{=} U_{k,k}^{(1)} U_{k+1,k+1}^{(2)} \cdots U_{n,n}^{(n-k+1)}, \tag{4.23}$$

where the multipliers

$$U_{r,r}^{(r-k+1)}, \quad r = k, \ldots, n,$$

on the RHS of (4.23) are the maximum order statistics of $(n - k + 1)$ independent samples from the uniform distributions, which sizes are $k, k+1, \ldots, n$ respectively.

Example 4.6. There is one more representation for the uniform order statistics. Let again $v_1, v_2, \ldots$ be independent random variables having the standard $E(1)$ exponential distribution and

$$S_n = v_1 + v_2 + \cdots + v_n, \quad n = 1, 2, \ldots.$$

It turns out that the following representation is valid for the uniform order statistics:

$$(U_{1,n}, \ldots, U_{n,n}) \stackrel{d}{=} \left(\frac{S_1}{S_{n+1}}, \ldots, \frac{S_n}{S_{n+1}} \right). \tag{4.24}$$

To prove (4.24) one must recall (see (2.10) with $f(x) = 1$, $0 < x < 1$) that the joint pdf of order statistics $(U_{1,n}, \ldots, U_{n,n})$ has the form:

$$f_{1,2,\ldots,n:n}(x_1, x_2, \ldots, x_n) = n!, \quad 0 < x_1 < x_2 < \cdots < x_n < 1, \tag{4.25}$$

and it equals zero otherwise.

Now we will prove that the pdf of the RHS of (4.24) coincides with (4.25). Consider the joint pdf of random variables $v_1, v_2, \ldots, v_{n+1}$, which naturally is of the form:

$$g(v_1, \ldots, v_{n+1}) = \exp\{-(v_1 + \cdots + v_{n+1})\}, \quad v_1 \geqslant 0, \ldots, v_{n+1} \geqslant 0. \tag{4.26}$$

Using the linear transformation

$$(y_1, y_2, \ldots, y_{n+1}) = \left(\frac{v_1}{v_1 + \cdots + v_{n+1}}, \ldots, \frac{v_1 + v_2 + \cdots + v_n}{v_1 + \cdots + v_{n+1}}, v_1 + \cdots + v_{n+1} \right),$$

one gets the joint distribution density $h(y_1, y_2, \ldots, y_n, y_{n+1})$ of a set of random variables

$$\frac{S_1}{S_{n+1}}, \ldots, \frac{S_n}{S_{n+1}}$$

and S_{n+1} as follows:

$$h(y_1, y_2, \ldots, y_n, y_{n+1}) = (y_{n+1})^n \exp(-y_{n+1}), \tag{4.27}$$

if $0 < y_1 < \cdots < y_n$, $y_{n+1} \geqslant 0$.

The next step, integration over y_{n+1}, gives us the final formula for the joint p.d.f. of ratios

$$\frac{S_1}{S_{n+1}}, \ldots, \frac{S_n}{S_{n+1}}.$$

It is not difficult to see that this expression coincides with (4.25). Moreover, from (4.27) we get also the independence of the vector

$$\left(\frac{S_1}{S_{n+1}}, \ldots, \frac{S_n}{S_{n+1}} \right)$$

and the sum S_{n+1}.

Representation (4.24) can be rewritten in the following useful form:

$$(U_{1,n}, U_{2,n} - U_{1,n}, \ldots, U_{n,n} - U_{n-1,n}) \stackrel{d}{=} \left(\frac{v_1}{v_1 + \cdots + v_{n+1}}, \ldots, \frac{v_n}{v_1 + \cdots + v_{n+1}} \right) \tag{4.28}$$

Exercise 4.7. Show that uniform quasi-midrange

$$U_{n-k+1,n} - U_{k,n}, \quad k \leqslant n/2$$

has the same distribution as $U_{n-2k+1,n}$.

Exercise 4.8. We know from (4.24) that

$$(U_{1,n}, \ldots, U_{n,n}) \stackrel{d}{=} \left(\frac{S_1}{S_{n+1}}, \ldots, \frac{S_n}{S_{n+1}} \right),$$

where

$$S_k = v_1 + v_2 + \cdots + v_k, \quad k = 1, 2, \ldots$$

and $v_1, v_2, \ldots$ are independent random variables having the standard $E(1)$ exponential distribution. There is one more representation of the uniform order statistics in terms of the sums S_k, namely,

$$(U_{1,n}, \ldots, U_{n,n}) \stackrel{d}{=} (S_1, \ldots, S_n \mid S_{n+1} = 1), \qquad (4.29)$$

i.e., the distribution of the vector of uniform order statistics coincides with the conditional distribution of the vector of sums $S_1, \ldots, S_n$ given that $S_{n+1} = 1$. Prove representation (4.29).

Example 4.7. Let a stick of the unit length be randomly broken in n places. We get $(n+1)$ pieces of the stick. Let us arrange these pieces of the stick in the increasing order with respect to their lengths. What is the distribution of the k-th item in this variational series? The given construction deals with two orderings. Coordinates of the break points coincide with the uniform order statistics $U_{1,n}, U_{2,n}, \ldots, U_{n,n}$. Hence, the lengths of random pieces are

$$\delta_1 = U_{1,n}, \ \delta_2 = U_{2,n} - U_{1,n}, \ldots, \ \delta_n = U_{n,n} - U_{n-1,n}, \ \delta_{n+1} = 1 - U_{n,n}.$$

Let

$$\delta_{1,n+1} \leqslant \delta_{2,n+1} \leqslant \cdots \leqslant \delta_{n+1,n+1}$$

denote order statistics based on $\delta_1, \delta_2, \ldots, \delta_{n+1}$. From (4.28) we find that

$$(\delta_1, \delta_2, \ldots, \delta_{n+1}) \stackrel{d}{=} \left(\frac{v_1}{v_1 + \cdots + v_{n+1}}, \ldots, \frac{v_{n+1}}{v_1 + \cdots + v_{n+1}} \right). \qquad (4.30)$$

The ordering of $\delta_1, \delta_2, \ldots, \delta_{n+1}$ is equivalent to ordering of the exponential random variables $v_1, v_2, \ldots, v_{n+1}$. Hence, we obtain that

$$(\delta_{1,n+1}, \delta_{2,n+1}, \ldots, \delta_{n+1,n+1}) \stackrel{d}{=} \left(\frac{v_{1,n+1}}{v_1 + \cdots + v_{n+1}}, \ldots, \frac{v_{n+1,n+1}}{v_1 + \cdots + v_{n+1}} \right). \qquad (4.31)$$

Recalling that

$$v_{1,n+1} + v_{2,n+1} + \cdots + v_{n+1,n+1} = v_1 + v_2 + \cdots + v_{n+1},$$

one can express the RHS of (4.31) as

$$\left(\frac{v_{1,n+1}}{v_{1,n+1} + \cdots + v_{n+1,n+1}}, \ldots, \frac{v_{n+1,n+1}}{v_{1,n+1} + \cdots + v_{n+1,n+1}} \right).$$

Now we can apply representation (4.15) to exponential order statistics

$$v_{1,n+1}, \ v_{2,n+1}, \ldots, \ v_{n+1,n+1}.$$

Making some natural changing in (4.15) one comes to the appropriate result for ordered lengths of the pieces of the broken stick. It turns out that

$$\delta_{k,n+1} \stackrel{d}{=} \left(\frac{v_1}{n+1} + \frac{v_2}{n} + \cdots + \frac{v_k}{n-k+2} \right) \Big/ (v_1 + \cdots + v_{n+1}),$$

$$k = 1, 2, \ldots, n+1, \qquad (4.32)$$

where $v_1, \ldots, v_{n+1}$ are independent random variables having the standard $E(1)$ exponential distribution. In particular,

$$\delta_{1,n+1} \stackrel{d}{=} \frac{v_1}{(n+1)(v_1 + \cdots + v_{n+1})}. \qquad (4.33)$$

This means that $(n+1)\delta_{1,n+1}$ has the same distribution as $U_{1,n}$.

Remark 4.6. Combining representations (4.6), (4.8), (4.15), (4.21), (4.24), (4.29) one can successfully express distributions of arbitrary order statistics $X_{k,n}$ (related to a some d.f. F) in terms of distributions for sums or products of independent random variables. For instance, if G is the inverse of F and $v_1, v_2, \ldots$ are independent exponentially $E(1)$ distributed random variables then

$$X_{k,n} \stackrel{d}{=} G\left(\frac{v_1 + \cdots + v_k}{v_1 + \cdots + v_{n+1}} \right)$$

$$\stackrel{d}{=} G\left(1 - \exp\left(-\left(\frac{v_1}{n} + \frac{v_2}{n-1} + \cdots + \frac{v_k}{n-k+1} \right) \right) \right), \quad k = 1, 2, \ldots, n. \quad (4.34)$$

Exercise 4.9. Let $X_{1,n}, \ldots, X_{n,n}$ be order statistics corresponding to the distribution with the density

$$f(x) = ax^{a-1}, \quad 0 < x < 1, \ a > 0.$$

Express $X_{r,n}$ and the product $X_{r,n}X_{s,n}$, $1 \leqslant r < s \leqslant n$, in terms of independent uniformly distributed random variables.

Check your solutions

Exercise 4.1 (solution). It is clear that $0 \leqslant Y \leqslant 1$. Fix any $z \in (0, 1)$. We have

$$P\{Y \leqslant z\} = P\{F(X) \leqslant z\}.$$

Since F is continuous, the equality $F(x) = z$ has one solution (denote it α_z) at least. There are two possible options:

1) F strongly increases at α_z. Then $G(z) = \alpha_z$. Evidently, in this case events

$$\{\omega : F(X(\omega)) \leqslant z\} \quad \text{and} \quad \{\omega : X(\omega) \leqslant G(z)\} = \{\omega : X(\omega) \leqslant \alpha_z\}$$

have the same probability. Noting that

$$P\{\omega : X(\omega) \leqslant \alpha_z\} = F(\alpha_z) = z,$$

one gets that

$$P\{Y \leqslant z\} = P\{X \leqslant G(z)\} = F(G(z)) = z.$$

2) α_z belongs to some constancy interval $[a, b]$ of d.f. F, where

$$a = \inf\{x : F(x) \geqslant z\} \quad \text{and} \quad b = \sup\{x : F(x) \leqslant z\}.$$

Then $F(a) = F(\alpha_z) = F(b) = z$ and $G(z) = a$.

Indeed, events $\{F(X) \leqslant z\}$ and $\{X \leqslant G(z)\} = \{X \leqslant a\}$ also have the same probability and

$$P\{Y \leqslant z\} = P\{X \leqslant G(z)\} = F(G(z)) = F(a) = F(\alpha_z) = z.$$

Thus, both necessary statements are proved.

Exercise 4.2 (answer).

$$G(x) = \begin{cases} 1, & \text{if } 0 < x \leqslant 1/3, \\ 2, & \text{if } 1/3 < x \leqslant 2/3, \text{ and} \\ 3, & \text{if } 2/3 < x < 1; \end{cases}$$

random variable $G(U)$ has the same distribution as X.

Exercise 4.3 (solution). Note that Z_1 and Z_2 have d.f.'s

$$F_1(x) = 1 - e^{-x/\lambda}, \quad x \geqslant 0,$$

and

$$F_2(x) = 1 - e^{-x/\mu}, \quad x \geqslant 0,$$

correspondingly. One can see that

$$P\{\min\{Z_1, Z_2\} > x, \max\{Z_1, Z_2\} \leqslant y\} = P\{x < Z_1 \leqslant y, \, x < Z_2 \leqslant y\}$$

$$= (F_1(y) - F_1(x))(F_2(y) - F_2(x)) = (e^{-x/\lambda} - e^{-y/\lambda})(e^{-x/\mu} - e^{-y/\mu}), \quad 0 < x < y.$$

By differentiating over x and y we get that the joint probability density function $g(x,y)$ of $\min\{Z_1, Z_2\}$ and $\max\{Z_1, Z_2\}$ has the form

$$g(x,y) = \left(\exp\left(-\frac{x}{\lambda} - \frac{y}{\mu}\right) + \exp\left(-\frac{x}{\mu} - \frac{y}{\lambda}\right)\right)\Big/\lambda\mu, \quad 0 < x < y.$$

Consider the linear change of variables $(v, w) = (x, y - x)$ with the unit Jacobian, which corresponds to the passage to random variables V and $W > 0$. This transformation gives us the joint pdf of random variables V and W:

$$f_{V,W}(v,w) = \left(\exp\left(-\frac{v}{\lambda} - \frac{v+w}{\mu}\right) + \exp\left(-\frac{v}{\mu} - \frac{v+w}{\lambda}\right)\right)\Big/\lambda\mu, \quad v > 0, \; w > 0.$$

One sees that

$$f_{V,W}(v,w) = \frac{h_1(v)h_2(w)}{\lambda\mu}, \quad v > 0, \; w > 0,$$

where

$$h_1(v) = \exp\left(-v\left(\frac{1}{\lambda} + \frac{1}{\mu}\right)\right)$$

and

$$h_2(w) = \left(\exp\left(-\frac{w}{\mu}\right) + \exp\left(-\frac{w}{\lambda}\right)\right).$$

The existence of the factorization

$$f_{V,W}(v,w) = \frac{h_1(v)h_2(w)}{\lambda\mu}$$

enables us to state that random variables V and W are independent. Moreover, we can see that the functions

$$r_1(v) = \frac{\lambda + \mu}{\lambda\mu} h_1(v) = \frac{\lambda + \mu}{\lambda\mu} \exp\left(-\frac{\lambda + \mu}{\lambda\mu} v\right), \quad v > 0,$$

and

$$r_2(w) = \frac{1}{\lambda + \mu} h_2(w) = \frac{1}{\lambda + \mu}\left(\exp\left(-\frac{w}{\mu}\right) + \exp\left(-\frac{w}{\lambda}\right)\right), \quad w > 0,$$

present here the probability density functions of random variables V and W respectively.

Exercise 4.4 (solution). It follows from (4.15) that

$$(Z_{1,n}, L) \overset{d}{=} \left(\frac{v_1}{n}, \sum_{k=2}^{n} b_k v_k \right),$$

where

$$b_k = \frac{1}{n-k+1} \sum_{j=k}^{n} c_j.$$

Since v_k, $k = 1, 2, \ldots, n$, are independent random variables,

$$\frac{v_1}{n} \quad \text{and} \quad \sum_{k=2}^{n} b_k v_k$$

are also independent. Hence, so are $Z_{1,n}$ and L.

Exercise 4.5 (solution). The statement of exercise 4.5 immediately follows from (4.20), so far as simple transformations enable us to get the equality

$$(V_1, V_2, \ldots, V_n) \overset{d}{=} (W_1, W_2, \ldots, W_n),$$

where W's are independent uniformly distributed random variables.

Exercise 4.6 (solution). It suffices to understand that m components

$$\left(W_1 W_2^{1/2} \cdots W_{n-1}^{1/(n-1)} W_n^{1/n}, \ldots, W_m^{1/m} \cdots W_{n-1}^{1/(n-1)} W_n^{1/n} \right)$$

of vector (4.20), which correspond to order statistics $(U_{1,n}, \ldots, U_{m,n})$, can be given in the form

$$\left(W_1 W_2^{1/2} \cdots W_{n-1}^{1/(n-1)} W_m^{1/m} T, \ldots, W_m^{1/m} T \right),$$

where

$$T = W_{m+1}^{1/(m+1)} \cdots W_{n-1}^{1/(n-1)} W_n^{1/n},$$

does not depend on $W_1, W_2, \ldots, W_m$. Noting that vector

$$\left(W_1 W_2^{1/2} \cdots W_{n-1}^{1/(n-1)} W_m^{1/m}, \ldots, W_m^{1/m} \right)$$

has the same distribution as $(U_{1,n}, \ldots, U_{m,n})$, while T is distributed as $V_{m+1,n}$, we complete the solution of the exercise.

Exercise 4.7 (solution). From (4.24) we obtain that $U_{n-k+1,n} - U_{k,n}$ has the same distribution as the ratio

$$\frac{v_{k+1} + v_{k+2} + \cdots + v_{n-k+1}}{v_1 + \cdots + v_{n+1}}.$$

It is evident that this ratio and ratio

$$\frac{v_1 + v_2 + \cdots + v_{n-2k+1}}{v_1 + \cdots + v_{n+1}},$$

which corresponds to $U_{n-2k+1,n}$, also have the same distribution.

Exercise 4.8 (solution). The joint pdf of random variables $v_1, v_2, \ldots, v_{n+1}$ is given as

$$f_{n+1}(x_1, x_2, \ldots, x_{n+1}) = \exp(-(x_1 + x_2 + \cdots + x_{n+1})), \quad x_1 > 0, \ x_2 > 0, \ldots, \ x_{n+1} > 0.$$

By means of the linear transformation

$$(y_1, y_2, \ldots, y_{n+1}) = (x_1, x_1 + x_2, \ldots, x_1 + x_2 + \cdots + x_{n+1})$$

with a unit Jacobian one can get the following joint pdf for sums $S_1, S_2, \ldots, S_{n+1}$:

$$g_{n+1}(y_1, y_2, \ldots, y_{n+1}) = \exp(-y_{n+1}), \quad 0 < y_1 < y_2 < \cdots < y_{n+1}.$$

It is well known that a sum S_{n+1} of $(n+1)$ independent exponentially $E(1)$ distributed terms has the gamma distribution with parameter $n+1$:

$$h_{n+1}(v) = \frac{v^n \exp(-v)}{n!}, \quad v > 0.$$

To obtain the conditional density function of sums $S_1, S_2, \ldots, S_n$ given that $S_{n+1} = 1$ one has to calculate the expression

$$\frac{g_{n+1}(y_1, y_2, \ldots, y_n, 1)}{h_{n+1}(1)},$$

which evidently equals $n!$, if $0 < y_1 < y_2 < \cdots < y_n$, and equals zero, otherwise. This expression coincides with the joint pdf of the uniform order statistics given in (4.25). It means that representation (4.29) is true.

Exercise 4.9 (solution). One obtains that the corresponding d.f. in this exercise has the form

$$F(x) = x^a, \quad 0 < x < 1.$$

Hence, the inverse function is given as

$$G(x) = x^{1/a}, \quad 0 < x < 1,$$

and relations

$$X_{r,n} \stackrel{d}{=} (U_{r,n})^{1/a}$$

and

$$X_{r,n} X_{s,n} \stackrel{d}{=} (U_{r,n} U_{s,n})^{1/a}$$

are valid. Now we use representation (4.21) and have the following equalities:

$$X_{r,n} \stackrel{d}{=} W_r^{1/ra} W_{r+1}^{1/(r+1)a} \cdots W_n^{1/na}, \quad r = 1, 2, \ldots, n,$$

and

$$X_{r,n} X_{s,n} \stackrel{d}{=} W_r^{1/ra} W_{r+1}^{1/(r+1)a} \cdots W_{s-1}^{1/(s-1)a} W_s^{2/sa} W_{s+1}^{2/(s+1)a} \cdots W_n^{2/na}, \quad 1 \leqslant r < s \leqslant n,$$

where $W_1, W_2, \ldots, W_n$ are independent random variables with a common uniform on $[0,1]$ distribution.

Chapter 5

Conditional Distributions of Order Statistics

The structure of conditional distributions of order statistics is discussed. For instance, we show that sets of order statistics $(X_{1,n}, \ldots, X_{r-1,n})$ and $(X_{r+1,n}, \ldots, X_{n,n})$ become conditionally independent if $X_{r,n}$ is fixed. The Markov property of order statistics is also under investigation. It turns out that order statistics form a Markov chain if the underlying distribution is continuous, while this property fails, as a rule, for distribution functions having discontinuity points.

There are some useful relations related to conditional distributions of order statistics.

Example 5.1. Let

$$X_{1,n} \leqslant \cdots \leqslant X_{r-1,n} \leqslant X_{r,n} \leqslant X_{r+1,n} \leqslant \cdots \leqslant X_{n,n}$$

be a variational series corresponding to a distribution with a density $f(x)$. Fix a value of $X_{r,n}$ and consider the conditional distribution of the rest elements of the variational series given that $X_{r,n} = v$. We suppose that $f_{r:n}(v) > 0$ for this value v, where $f_{r:n}(v)$, as usual, denotes the pdf of $X_{r,n}$. Let

$$f_{1,\ldots,r-1,r+1,\ldots,n|r}(x_1, \ldots, x_{r-1}, x_{r+1}, \ldots, x_n \mid v)$$

denote the joint conditional density of order statistics

$$X_{1,n}, \ldots, X_{r-1,n}, X_{r+1,n}, \ldots, X_{n,n}$$

given that $X_{r,n} = v$. The definition of the conditional densities requires to know the (unconditional) joint pdf of all random variables (including fixed) and separately the pdf of the set of fixed random variables. In our case these pdf's are $f_{1,2,\ldots,n:n}$, the joint pdf of all order statistics $X_{1,n}, \ldots, X_{n,n}$, and $f_{r:n}$, the pdf of $X_{r:n}$. The standard procedure gives us the required density function:

$$f_{1,\ldots,r-1,r+1,\ldots,n|r}(x_1, \ldots, x_{r-1}, x_{r+1}, \ldots, x_n \mid v)$$
$$= \frac{f_{1,2,\ldots,n:n}(x_1, \ldots, x_{r-1}, v, x_{r+1}, \ldots, x_n)}{f_{r:n}(v)}. \tag{5.1}$$

M. Ahsanullah et al., *An Introduction to Order Statistics*,
Atlantis Studies in Probability and Statistics 3,
DOI: 10.2991/978-94-91216-83-1_5, © Atlantis Press 2013

Upon substituting (2.10) and (2.13) in (5.1), we get that

$$f_{1,\ldots,r-1,r+1,\ldots,n|r}(x_1,\ldots,x_{r-1},x_{r+1},\ldots,x_n \mid v)$$

$$= (r-1)!\prod_{k=1}^{r-1}\frac{f(x_k)}{F(v)}(n-r)!\prod_{k=r+1}^{n}\frac{f(x_k)}{1-F(v)}, \qquad (5.2)$$

if $x_1 < \cdots < x_{r-1} < v < x_{r+1} < \cdots < x_n$, and the LHS of (5.2) equals zero otherwise.
For each value v we introduce d.f.'s

$$G(x,v) = P\{X \leqslant x \mid X \leqslant v\} = \frac{F(x)}{F(v)}, \quad x \leqslant v, \qquad (5.3)$$

and

$$H(x,v) = P\{X \leqslant x \mid X > v\} = \frac{F(x)-F(v)}{1-F(v)}, \quad x > v. \qquad (5.4)$$

The corresponding densities have the form

$$g(x,v) = \begin{cases} \dfrac{f(x)}{F(v)}, & \text{if } x \leqslant v, \\ 0, & \text{if } x > v; \end{cases} \qquad (5.5)$$

and

$$h(x,v) = \begin{cases} \dfrac{f(x)}{1-F(v)}, & \text{if } x > v, \\ 0, & \text{if } x \leqslant v. \end{cases} \qquad (5.6)$$

Now (5.2) can be expressed as follows:

$$f_{1,\ldots,r-1,r+1,\ldots,n|r}(x_1,\ldots,x_{r-1},x_{r+1},\ldots,x_n \mid v) = g(x_1,\ldots,x_{r-1},v)h(x_{r+1},\ldots,x_n,v), \qquad (5.7)$$

where

$$g(x_1,\ldots,x_{r-1},v) = \begin{cases} (r-1)!g(x_1,v)\cdots g(x_{r-1},v), & \text{if } x_1 < \cdots < x_{r-1} < v, \\ 0, & \text{otherwise,} \end{cases}$$

while

$$h(x_{r+1},\ldots,x_n) = \begin{cases} (n-r)!g(x_{k+1},v)\cdots g(x_n,v), & \text{if } v < x_{r+1} < \cdots < x_n, \\ 0, & \text{otherwise.} \end{cases}$$

We immediately derive from (5.7) that two sets of order statistics,

$$(X_{1,n},\ldots,X_{r-1,n}) \text{ and } (X_{r+1,n},\ldots,X_{n,n})$$

are conditionally independent given any fixed value of $X_{r,n}$. Observing the form of $g(x_1,\ldots,x_{r-1},v)$ one can see that this conditional joint pdf, corresponding to order statistics $X_{1,n},\ldots,X_{r-1,n}$, coincides with unconditional joint pdf of order statistics, say, $Y_{1,r-1} \leqslant \cdots \leqslant Y_{r-1,r-1}$, corresponding to a population with d.f. $G(x,v)$ and density $g(x,v)$. This assertion we will write in the following form, where $F(x)$ and $G(x,v)$ denote population distribution functions:

$$\{F(x);X_{1,n},\ldots,X_{r-1,n} \mid X_{r,n}=v\} \overset{d}{=} \{G(x,v);Y_{1,r-1},\ldots,Y_{r-1,r-1}\}. \qquad (5.8)$$

Similarly we obtain that the conditional distribution of order statistics $X_{r+1,n}, \ldots, X_{n,n}$ given that $X_{r,n} = v$ coincides with the unconditional distribution of order statistics, say, $W_{1,n-r} \leqslant \cdots \leqslant W_{n-r,n-r}$ related to d.f. $H(x,v)$ and pdf $h(x,v)$:

$$\{F(x); X_{r+1,n}, \ldots, X_{n,n} \mid X_{r,n} = v\} \overset{d}{=} \{H(x,v); W_{1,n-r}, \ldots, W_{n-r,n-r}\}. \tag{5.9}$$

Let $Y_1, Y_2, \ldots, Y_{r-1}$ be a sample of size $(r-1)$ from a population with d.f. $G(x,v)$. Then the following corollary of (5.8) is valid:

$$\{F(x); X_{1,n} + \cdots + X_{r-1,n} \mid X_{r,n} = v\} \overset{d}{=} \{G(x,v); Y_{1,r-1} + \cdots + Y_{r-1,r-1}\}$$

$$\overset{d}{=} \{G(x,v); Y_1 + \cdots + Y_{r-1}\}. \tag{5.10}$$

Relation (5.10) enable us to express the distribution of the sum $X_{1,n} + \cdots + X_{r-1,n}$ as a mixture (taken over the parameter v) of $(r-1)$-fold convolutions of d.f.'s $G(x,v)$. Indeed, the similar corollary is valid for the sum $X_{r+1,n} + \cdots + X_{n,n}$.

Exercise 5.1. What is the structure of the conditional distribution of order statistics

$$X_{1,n}, \ldots, X_{r-1,n}; X_{r+1,n}, \ldots, X_{s-1,n} \text{ and } X_{s+1,n}, \ldots, X_{n,n},$$

given that two order statistics $X_{r,n} = v$ and $X_{s,n} = z$, $r < s$, $v < z$ are fixed?

Exercise 5.2. Show that the conditional distribution of the uniform order statistics $(U_{1,n}, \ldots, U_{k,n})$ given $U_{k+1,n} = v$ coincides with the unconditional distribution of the vector $(vU_{1,k}, \ldots, vU_{k,k})$.

Exercise 5.3. Show that the conditional distribution of the uniform quasi-range $U_{n-k+1,n} - U_{k,n}$ given that $U_{n,n} - U_{1,n} = u$ coincides with the unconditional distribution of $u(U_{n-k,n-2} - U_{k-1,n-2})$.

Exercise 5.4. Prove that the conditional distribution of the exponential order statistics $(Z_{r+1,n}, \ldots, Z_{n,n})$ given $Z_{r,n} = v$ coincides with the unconditional distribution of the vector $(v + Z_{1,n-r}, \ldots, v + Z_{n-r,n-r})$.

Remark 5.1. The assertion obtained in example 5.1 stays valid for any continuous distribution function F. The following exercise shows that the conditional independence property can fail if d.f. F has jump points.

Exercise 5.5. Let X have an atom $P\{X = a\} = p > 0$ at some point a. Let also

$$P\{X < a\} = F(a) - p > 0 \text{ and } P\{X > a\} = 1 - F(a) > 0.$$

Consider order statistics $X_{1,3} \leqslant X_{2,3} \leqslant X_{3,3}$. Show that $X_{1,3}$ and $X_{3,3}$ are conditionally dependent given that $X_{2,3} = a$.

Remark 5.2. We see that conditional independence of order statistics

$$X_{1,n} \leqslant \cdots \leqslant X_{r-1,n} \text{ and } X_{r+1,n} \leqslant \cdots \leqslant X_{n,n},$$

given that $X_{r,n}$ is fixed, can fail if there is a positive probability that two neighboring elements of the variational series coincide. The conditional independence property can be saved under additional condition that $X_{1,n} < X_{2,n} < \cdots < X_{n,n}$. The next exercise illustrates this fact.

Exercise 5.6. Let X take on values $x_1, x_2, \ldots$ with non-zero probabilities and let $X_{1,n}$, $X_{2,n}, \ldots, X_{n,n}$ be the corresponding order statistics. We suppose that a number of possible values of X is not less than n. Show that the vectors

$$(X_{1,n}, \ldots, X_{r-1,n}) \text{ and } (X_{r+1,n}, \ldots, X_{n,n})$$

are conditionally independent given that $X_{r,n} = v$ is fixed and $X_{1,n} < X_{2,n} < \cdots < X_{n,n}$. Indeed, the value v is chosen to satisfy the condition

$$P\{X_{1,n} < \cdots < X_{r-1,n} < X_{r,n} = v < X_{r+1,n} < \cdots < X_{n,n}\} > 0.$$

Now we will discuss the Markov property of order statistics. We consider a sequence of order statistics $X_{1,n}, X_{2,n}, \ldots, X_{n,n}$, corresponding to a population with a density function f.

Example 5.2. From (5.7) we found that under fixed value of $X_{r,n}$ order statistics $X_{r+1,n}, \ldots, X_{n,n}$ are conditionally independent on random variables $X_{1,n}, \ldots, X_{r-1,n}$. The same arguments are used to check the Markov property of order statistics. We need to prove that for any $r = 3, 4, \ldots, n$, the conditional density

$$f_{r|1,2,\ldots,r-1}(u \mid x_1, \ldots, x_{r-1})$$

of $X_{r,n}$ given that $X_{1,n} = x_1, \ldots, X_{r-1,n} = x_{r-1}$, coincides with the conditional density $f_{r|r-1}(u \mid x_{r-1})$ of $X_{r,n}$ given only that $X_{r-1,n} = x_{r-1}$ is fixed. Recalling the definitions of conditional densities, one finds that

$$
\begin{aligned}
f_{r|1,2,\ldots,r-1}(u \mid x_1, \ldots, x_{r-1}) &= \frac{f_{1,2,\ldots,r-1,r:n}(x_1, x_2, \ldots, x_{r-1}, u)}{f_{1,2,\ldots,r-1:n}(x_1, x_2, \ldots, x_{r-1})} \\
&= \frac{f_{1,2,\ldots,r-1|r}(x_1, x_2, \ldots, x_{r-1} \mid u) f_{r:n}(u)}{f_{1,2,\ldots,r-2|r-1}(x_1, x_2, \ldots, x_{r-1} \mid x_{r-1}) f_{r-1,n}(x_{r-1})}. \quad (5.11)
\end{aligned}
$$

In (5.11) $f_{1,2,\ldots,r-1|r}(x_1, x_2, \ldots, x_{r-1} \mid u)$ is the joint conditional density of $X_{1,n}, \ldots, X_{r-1,n}$ given that $X_{r,n} = u$, and it coincides, as we know, with

$$g(x_1, \ldots, x_{r-1}, u) = (r-1)! g(x_1, u) \cdots g(x_{r-1}, u)$$

determined in (5.7), where

$$g(x,u) = \frac{f(x)}{F(u)}, \quad x < u.$$

From (2.13) we also know that

$$f_{r:n}(x) = \frac{n!}{(r-1)!(n-r)!}(F(x))^{r-1}(1-F(x))^{n-r}f(x).$$

The similar expressions (by changing r for $r-1$) can be written for the rest terms on the RHS of (5.11). Substituting all these expressions in (5.11) one obtains that

$$f_{r|1,2,\ldots,r-1}(u \mid x_1,\ldots,x_{r-1}) = \frac{(n-r+1)(1-F(u))^{n-r}f(u)}{(1-F(x_{r-1}))^{n-r+1}}, \quad u > x_{r-1}. \tag{5.12}$$

It suffices to obtain that

$$f_{r|r-1}(u \mid x_{r-1}) = \frac{f_{r-1,r:n}(x_{r-1},u)}{f_{r-1:n}(x_{r-1})}$$

coincides with the expression on the RHS of (5.12). Due to relations (2.14) and (2.15), which give us density functions $f_{r-1:n}(x_{r-1})$ and $f_{r-1,r:n}(x_{r-1},u)$, we easily prove the desired statement.

Exercise 5.7. Give another proof of (5.11), based on the equality

$$f_{r|1,2,\ldots,r-1}(u \mid x_1,\ldots,x_{r-1}) = \frac{f_{1,2,\ldots,r-1,r:n}(x_1,x_2,\ldots,x_{r-1},u)}{f_{1,2,\ldots,r-1:n}(x_1,x_2,\ldots,x_{r-1})}.$$

Remark 5.3. It follows from example 5.2 that a sequence $X_{1,n},\ldots,X_{n,n}$ forms a Markov chain and

$$P\{X_{k+1,n} > x \mid X_{k,n} = u\} = \left(\frac{1-F(x)}{1-F(u)}\right)^{n-k}, \quad x > u, \tag{5.13}$$

for any $k = 1, 2, \ldots, n-1$. Note also that we assumed that the underlying distribution has a density function only for the sake of simplicity. In fact, order statistics satisfies the Markov property in a more general situation, when a population has any continuous d.f. F, while this property can fail if F has some jump points.

Example 5.3. In exercise 5.5 we considered a distribution having an atom $P\{X = a\} = p > 0$ at some point a and supposed that $b = P\{X < a\} = F(a) - p > 0$ and $c = P\{X > a\} = 1 - F(a) > 0$. It turns out in this situation that $X_{1,3}$ and $X_{3,3}$ are conditionally dependent given that $X_{2,3} = a$. We can propose that order statistics $X_{1,3}$, $X_{2,3}$ and $X_{3,3}$ can not possess the Markov structure for such distributions. In fact, in this case

$$P\{X_{1,3} = a, X_{2,3} = a\} = p^3 + 3p^2c, \tag{5.14}$$

$$P\{X_{1,3} = a, X_{2,3} = a, X_{3,3} = a\} = p^3, \tag{5.15}$$

$$P\{X_{2,3} = a, X_{3,3} = a\} = p^3 + 3p^2b \tag{5.16}$$

and

$$P\{X_{2,3} = a\} = p^3 + 3p^2(1-p) + 6pbc. \tag{5.17}$$

From (5.13) and (5.14) we obtain that

$$P\{X_{3,3} = a \mid X_{2,3} = a, X_{1,3} = a\} = \frac{p}{p+3c}, \tag{5.18}$$

while, due to (5.15) and (5.16),

$$P\{X_{3,3} = a \mid X_{2,3} = a\} = \frac{p^2 + 3pb}{p^2 + 3p(1-p) + 6bc}. \tag{5.19}$$

It is not difficult to check now that equality

$$P\{X_{3,3} = a \mid X_{2,3} = a, X_{1,3} = a\} = P\{X_{3,3} = a \mid X_{2,3} = a\}$$

is equivalent to the relation $bc = 0$, which fails so far as $b > 0$ and $c > 0$ for considered distributions.

Remark 5.4. It follows from example 5.3 that the Markov property is not valid for order statistics if the underlying d.f. has three jump points or more, because in this situation there exists one point a at least such that $P\{X = a\} > 0$, $P\{X < a\} > 0$ and $P\{X > a\} > 0$. Hence it remains to discuss distributions with one or two atoms, which coincide with end-points $\alpha = \inf\{x : F(x) > 0\}$ or/and $\beta = \sup\{x : F(x) < 1\}$ of the distribution support.

Example 5.4. Let X have degenerate distribution, say $P\{X = a\} = 1$. Then it is evident that order statistics form the Markov property.

Exercise 5.8. Let X take on two values $a < b$ with probabilities

$$0 < p = P\{X = a\} = 1 - P\{X = b\} < 1.$$

Show that order statistics $X_{1,n}, \ldots, X_{n,n}$ form a Markov chain.

Remark 5.5. It seems that the following hypotheses must be true.
a) Let the d.f. F have one discontinuity point only, which coincides with the left end-point $\alpha = \inf\{x : F(x) > 0\} > -\infty$ or with the right end-point $\beta = \sup\{x : F(x) < 1\} < \infty$ of the distribution support. Then order statistics $X_{1,n}$, $X_{2,n}$ and $X_{n,n}$ possess the Markov property.
b) If α and β are the only jump points of d.f. F, then order statistics form a Markov chain. We suggest our readers to check the validity of these hypotheses or to construct counter-examples, which reject them.

Example 5.5. Let $X_{1,n} \leqslant \cdots \leqslant X_{r-1,n} \leqslant X_{r,n} \leqslant X_{r+1,n} \leqslant \cdots \leqslant X_{n,n}$ be a variational series corresponding to a distribution with the density $f(x) = e^{-x}$, $x \geqslant 0$. Then the conditional pdf of $X_{r+1,n} \mid X_{1,n}, \ldots, X_{r,n})$ is

$$f_{r+1|1,2,\ldots,r}(x_{r+1} \mid X_{1,n} = x_1, \ldots, X_{r,n} = x_r) = (n-r)e^{-(n-r)(x_{r+1}-x_r)}, \quad 0 \leqslant x_r \leqslant x_{r+1} < \infty.$$

If $D_r = (n-r+1)(X_{r,n} - X_{r-1,n})$, then D_r, $r = 1, 2, \ldots, n$ with $X_{0,n} = 0$, are independent and identically distributed as exponential with cdf $F(x) = 1 - e^{-x}$, $x \geqslant 0$.

Check your solutions

Exercise 5.1 (answer). We have three conditionally independent sets of order statistics. Analogously to the case from example 5.1 the following relations (similar to (5.9) and (5.10)) are valid:

$$\{F(x); X_{1,n}, \ldots, X_{r-1,n} \mid X_{r,n} = v, X_{s,n} = z\} \overset{d}{=} \{G(x,v); Y_{1,r-1}, \ldots, Y_{r-1,r-1}\},$$

$$\{F(x); X_{r+1,n}, \ldots, X_{s-1,n} \mid X_{r,n} = v, X_{s,n} = z\} \overset{d}{=} \{T(x,v,z); V_{1,s-r-1}, \ldots, V_{s-r-1,s-r-1}\}$$

and

$$\{F(x); X_{s+1,n}, \ldots, X_{n-s,n} \mid X_{r,n} = v, X_{s,n} = z\} \overset{d}{=} \{H(x,z); W_{1,n-s}, \ldots, W_{n-s,n-s}\},$$

where $Y_{1,r-1}, \ldots, Y_{r-1,r-1}$ correspond to d.f.

$$G(x,v) = P\{X \leqslant x \mid X \leqslant v\} = \frac{F(x)}{F(v)}, \quad x \leqslant v;$$

$W_{1,n-s}, \ldots, W_{n-s,n-s}$ are order statistics related to d.f.

$$H(x,z) = P\{X \leqslant x \mid X > z\} = \frac{F(x) - F(z)}{1 - F(z)}, \quad x > z,$$

and order statistics $V_{1,s-r-1} \leqslant \cdots \leqslant V_{s-r-1,s-r-1}$ correspond to d.f.

$$T(x,v,z) = P\{X \leqslant x \mid v < X \leqslant z\} = \frac{F(x) - F(v)}{F(z) - F(v)}, \quad v < x < z.$$

Exercise 5.2 (hint). It suffices to use relation (5.8) with the d.f.

$$G(x,v) = \frac{x}{v}, \quad 0 < x < v,$$

which corresponds to the random variable vU, where U has the standard uniform distribution.

Exercise 5.3 (hint). Use the statements and results of exercises 5.1 and 5.2. Consider the conditional distribution of the uniform order statistics $U_{n-k+1,n}$ and $U_{k,n}$ given that $U_{1,n} = v$ and $U_{n,n} = z$. You will find that

$$(U_{k,n}, U_{n-k+1,n} \mid U_{1,n} = v, U_{n,n} = z) \overset{d}{=} ((z-v)U_{k-1,n-2}, (z-v)U_{n-k,n-2})$$

and hence

$$(U_{n-k+1,n} - U_{k,n} \mid U_{1,n} = v, U_{n,n} = z) \overset{d}{=} (z-v)(U_{n-k,n-2} - U_{k-1,n-2}).$$

Note that really the desired conditional distribution depends on the difference $z - v$ rather than on z and v. Due to this fact one can show that

$$(U_{n-k+1,n} - U_{k,n} \mid U_{n,n} - U_{1,n} = u) \overset{d}{=} u(U_{n-k,n-2} - U_{k-1,n-2}).$$

Exercise 5.4 (hint). Show that in this situation the d.f. $H(x,v)$ from (5.9) correspond to the random variable $Z + v$, where Z has the exponential $E(1)$ distribution.

Exercise 5.5 (solution). Consider conditional probabilities

$$p_1 = P\{X_{1,3} = a, X_{3,3} = a \mid X_{2,3} = a\} = \frac{P\{X_{1,3} = a, X_{2,3} = a, X_{3,3} = a\}}{P\{X_{2,3} = a\}}$$

$$= \frac{P\{X_1 = a, X_2 = a, X_3 = a\}}{P\{X_{2,3} = a\}} = \frac{p^3}{P\{X_{2,3} = a\}},$$

$$p_2 = P\{X_{1,3} = a \mid X_{2,3} = a\} = \frac{P\{X_{1,3} = a, X_{2,3} = a\}}{P\{X_{2,3} = a\}} = \frac{p^3 + 3p^2(1 - F(a))}{P\{X_{2,3} = a\}}$$

and

$$p_3 = P\{X_{3,3} = a \mid X_{2,3} = a\} = \frac{P\{X_{2,3} = a, X_{3,3} = a\}}{P\{X_{2,3} = a\}} = \frac{p^3 + 3p^2(F(a) - p)}{P\{X_{2,3} = a\}}.$$

To provide the conditional independence of $X_{1,3}$ and $X_{3,3}$ we need at least to have the equality $p_1 = p_2 p_3$, which is equivalent to the following relation:

$$(p^3 + 3p^2(1 - F(a)))(p^3 + 3p^2(F(a) - p)) = p^3 P\{X_{2,3} = a\}.$$

Note also that

$$P\{X_{2,3} = a\} = p^3 + 3p^2 F(a-0) + 3p^2(1 - F(a)) + 6pF(a-0)(1 - F(a))$$

$$= p^3 + 3p^2(1 - p) + 6p(F(a) - p)(1 - F(a))$$

$$= 3p^2 - 2p^3 + 6p(F(a) - p)(1 - F(a))$$

and hence, the desired equality must have the form

$$(p^3 + 3p^2(1 - F(a)))(p^3 + 3p^2(F(a) - p)) = p^3(3p^2 - 2p^3 + 6p(F(a) - p)(1 - F(a))).$$

After some natural simplifications one sees that the latter equality is valid only if

$$(F(a) - p)(1 - F(a)) = 0,$$

but both possible solutions $F(a) - p = 0$ and $1 - F(a) = 0$ are rejected by restrictions of the exercise. Hence, $X_{1,3}$ and $X_{3,3}$ are conditionally dependent given that $X_{2,3} = a$.

Exercise 5.6 (solution). Let $v_1, v_2, \ldots, v_{r-1}, v, v_{r+1}, \ldots, v_n$ be any n values taken from a set $\{x_1, x_2, \ldots\}$. Note that

$$I(v_1, \ldots, v_{r-1}, v, v_{r+1}, \ldots, v_n)$$

$$= P\{X_{1,n} = v_1, \ldots, X_{r-1,n} = v_{r-1}, X_{r+1,n} = v_{r+1}, \ldots, X_{n,n} = v_n | X_{r,n} = v, X_{1,n} < X_{2,n} < \cdots < X_{n,n}\}$$

$$= \frac{P\{X_{1,n} = v_1, \ldots, X_{r-1,n} = v_{r-1}, X_{r,n} = v, X_{r+1,n} = x_{r+1}, \ldots, X_{n,n} = x_n, X_{1,n} < X_{2,n} < \cdots < X_{n,n}\}}{P\{X_{r,n} = v, X_{1,n} < X_{2,n} < \cdots < X_{n,n}\}}.$$

It is not difficult to see that

$$P\{X_{1,n} = v_1, \ldots, X_{r-1,n} = v_{r-1}, X_{r,n} = v, X_{r+1,n} = x_{r+1}, \ldots, X_{n,n} = x_n, X_{1,n} < X_{2,n} < \cdots < X_{n,n}\}$$

$$= n! P\{X = v\} \prod_{k=1}^{r-1} P\{X = v_k\} \prod_{k=r+1}^{n} P\{X = v_k\},$$

if $v_1 < v_2 < \cdots < v_{r-1} < v < v_{r+1} < \cdots < v_n$, and this probability equals zero otherwise. Then,

$$P\{X_{r,n} = v, X_{1,n} < X_{2,n} < \cdots < X_{n,n}\} = n! P\{X_1 < \cdots < X_{r-1} < X_r = v < X_{r+1} < \cdots < X_n\}$$

$$= n! P\{X = v\} P\{X_1 < \cdots < X_{r-1} < X_r = v\} P\{v < X_{r+1} < \cdots < X_n\}.$$

Thus,

$$I(v_1, \ldots, v_{r-1}, v, v_{r+1}, \ldots, v_n) = G(v_1, \ldots, v_{r-1}, v) H(v, v_{r+1}, \ldots, v_n),$$

where

$$G(v_1, \ldots, v_{r-1}, v) = \left(\prod_{k=1}^{r-1} P\{X = v_k\} \Big/ P\{X_1 < \cdots < X_{r-1} < X_r = v\} \right),$$

if $v_1 < v_2 < \cdots < v_{r-1} < v$, and

$$H(v, v_{r+1}, \ldots, v_n) = \left(\prod_{k=r+1}^{n} P\{X = v_k\} \Big/ P\{v < X_{r+1} < \cdots < X_n\} \right),$$

if $v < v_{r+1} < \cdots < v_n$.

The existence of the factorization

$$I(v_1, \ldots, v_{r-1}, v, v_{r+1}, \ldots, v_n) = G(v_1, \ldots, v_{r-1}, v) H(v, v_{r+1}, \ldots, v_n)$$

provides the conditional independence of order statistics.

Exercise 5.7 (hint). The expression for joint density functions

$$f_{1,2,\ldots,k}(x_1, x_2, \ldots, x_k), \quad k = r-1, r$$

present a particular case of the equalities, which were obtained in exercise 2.7.

Exercise 5.8 (hint). Fix any $2 < r \leqslant n$ and compare probabilities

$$P\{X_{r,n} = x_r \mid X_{r-1,n} = x_{r-1}, \ldots, X_{1,n} = x_1\} \text{ and } P\{X_{r,n} = x_r \mid X_{r-1,n} = x_{r-1}\}.$$

Since $x_1, x_2, \ldots, x_r$ is a non-decreasing sequence of zeros and ones, we need to consider only two situations. If $x_{r-1} = 0$, then the events

$$\{X_{r-1,n} = 0\} \text{ and } \{X_{r-1,n} = 0, \ldots, X_{1,n} = 0\}$$

coincide and then both conditional probabilities are equal. If $x_{r-1} = 1$, then $X_{r,n}$ is obliged to be equal 1 with probability one and both conditional probabilities are equal to one.

Chapter 6

Order Statistics for Discrete Distributions

Some properties of order statistics based on discrete distributions are discussed. Conditional and unconditional distributions of discrete order statistics are found. Some problems are solved for order statistics, which are based on geometric distributions. We also consider sampling without replacement from finite populations and consider the distributions and properties of the corresponding order statistics.

We consider order statistics $X_{1,n} \leqslant X_{2,n} \leqslant \cdots \leqslant X_{n,n}$ based on a sample $X_1, X_2, \ldots, X_n$ for the case, when a population distribution is discrete. Let $X_1, X_2, \ldots, X_n$ be n independent observations on random variable X taking on values $x_1, x_2, \ldots$ with probabilities $p_1, p_2, \ldots$.

Exercise 6.1. Show that

$$0 \leqslant p^2 \leqslant P\{X_{1,2} = X_{2,2}\} \leqslant p,$$

where

$$p = \max\{p_1, p_2, \ldots\}.$$

Exercise 6.2. Let X take on values $1, 2, \ldots, n$ with probabilities $p_r = 1/n$, $1 \leqslant r \leqslant n$. Find probabilities

$$\pi_r = P\{X_{1,r} < X_{2,r} < \cdots < X_{r,r}\}$$

for any r, $2 \leqslant r \leqslant n$.

Exercise 6.3. Let X take on two values, 0 and 1, with probabilities p and $g = 1 - p$ correspondingly. Find the distribution of $X_{k,n}$.

Exercise 6.4. In the previous exercise find the joint distribution of order statistics $X_{j,n}$ and $X_{k,n}$, $1 \leqslant j < k \leqslant n$, and the distribution of the difference

$$W_{j,k,n} = X_{k,n} - X_{j,n}.$$

M. Ahsanullah et al., *An Introduction to Order Statistics*,
Atlantis Studies in Probability and Statistics 3,
DOI: 10.2991/978-94-91216-83-1_6, © Atlantis Press 2013

Exercise 6.5. Let X have a geometric distribution with probabilities

$$P\{X = k\} = (1 - p)p^k, \quad k = 0, 1, 2, \ldots,$$

where $0 < p < 1$. Find distributions of order statistics $X_{1,n}$ and $X_{n,n}$.

Exercise 6.6. Under conditions of exercise 6.3 find the distribution of $X_{2,n}$.

Exercise 6.7. Find the distribution of

$$Y = \min\{Y_1, Y_2, \ldots, Y_n\},$$

where Y's are independent and have different geometric distributions with probabilities

$$P\{Y_r = k\} = (1 - p_r)p_r^k, \quad k = 0, 1, \ldots; \quad r = 1, 2, \ldots, n.$$

Exercise 6.8. Consider the case, when X takes on values $0, 1, 2, \ldots$ with probabilities $p_0, p_1, p_2, \ldots$. Find expressions for

$$P\{X_{k,n} = r\}, \quad k = 1, 2, \ldots, n; \quad r = 0, 1, \ldots.$$

Exercise 6.9. Let X have a geometric distribution with probabilities

$$p_k = P\{X = k\} = (1 - p)p^k, \quad k = 0, 1, 2, \ldots, \quad 0 < p < 1.$$

Find the common distribution of $X_{1,n}$ and $X_{n,n}$.

Exercise 6.10. Under conditions of exercise 6.9 find the distribution of sample ranges

$$W_n = X_{n,n} - X_{1,n}, \quad n = 2, 3, \ldots.$$

Exercise 6.11. We again consider the geometric distribution from exercises 6.9 and 6.10. Show that then $X_{1,n}$ and W_n are independent for $n = 2, 3, \ldots$.

Now we consider conditional distributions of discrete order statistics.

Exercise 6.12. Let X take on values $x_1 < x_2 < \cdots$ with probabilities $p_k = P\{X = k\}$. Find

$$P\{X_{1,3} = x_r, X_{3,3} = x_s \mid X_{2,3} = x_k\},$$

$$P\{X_{1,3} = x_r \mid X_{2,3} = x_k\}$$

and

$$P\{X_{3,3} = x_s \mid X_{2,3} = x_k\}, \quad x_r < x_k < x_s.$$

Remark 6.1. It is interesting to check when order statistics $X_{1,3}$ and $X_{3,3}$ are conditionally independent given that $X_{2,3}$ is fixed. In exercise 5.5 we have got that $X_{1,3}$ and $X_{3,3}$ are conditionally dependent for any distribution, which has an atom in some point a $(P\{X = a\} = p > 0)$, such that

$$P\{X < a\} = F(a) - p > 0$$

and

$$P\{X > a\} = 1 - F(a) > 0.$$

It means that for any discrete distribution, which has three or more values, order statistics $X_{1,3}$ and $X_{3,3}$ are conditionally dependent. Indeed, if X is degenerate and $P\{X = a\} = 1$, then

$$P\{X_{1,3} = a, X_{3,3} = a \mid X_{2,3} = a\} = 1 = P\{X_{1,3} = a \mid X_{2,3} = a\}P\{X_{3,3} = a \mid X_{2,3} = a\}$$

and hence $X_{1,3}$ and $X_{3,3}$ are conditionally independent given that $X_{2,3}$ is fixed. The only case, which we need to investigate now is the situation, when X takes on two values.

Exercise 6.13. Let

$$P\{X = a\} = 1 - P\{X = b\} = p,$$

where $0 < p < 1$ and $a < b$. Prove that $X_{1,3}$ and $X_{3,3}$ are conditionally independent given that $X_{2,3}$ is fixed.

Remark 6.2. In chapter 5 we investigated Markov properties of order statistics and found that they do not possess the Markov structure if X has three jump points or more. It was also shown that order statistics form a Markov chain if X is degenerate or if it takes on two values only.

Very often in statistics one uses sampling without replacement from a finite population. Let us have a set of N ordered distinct population values $x_1 < x_2 < \cdots < x_n$. If a sample of size n $(n \leqslant N)$ is drawn at random without replacement we deal with some dependent identically distributed random variables $(X_1, X_2, \ldots, X_n)$. The common distribution of these random variables are given by the equality

$$P\{X_1 = x_{k(1)}, X_2 = x_{k(2)}, \ldots, X_n = x_{k(n)}\} = \frac{1}{N(N-1)\cdots(N-n+1)}, \qquad (6.1)$$

which holds for any group of n distinct values $x_{k(1)}, x_{k(2)}, \ldots, x_{k(n)}$ taken from the original set of values $x_1, x_2, \ldots, x_n$. By arranging X's in increasing order we come to the corresponding order statistics

$$X_{1,n,N} < X_{2,n,X} < \cdots < X_{n,n,N},$$

where n denotes the sample size and N is the population size. We see now from (6.1) that

$$P\{X_{1,n,N} = x_{r(1)}, X_{2,n,N} = x_{r(2)}, \ldots, X_{n,n,N} = x_{r(n)}\} = \frac{n!}{N(N-1)\cdots(N-n+1)}$$

$$= \frac{1}{\binom{N}{n}}, \qquad (6.2)$$

where

$$x_{r(1)} < x_{r(2)} < \cdots < x_{r(n)}$$

are the ordered values $x_{k(1)}, x_{k(2)}, \ldots, x_{k(n)}$.

Simple combinatorial methods enable us to find different distributions of these order statistics.

Example 6.1. Let us find probabilities

$$p_{k,r} = P\{X_{k,n,N} = x_r\}$$

for $r = 1, 2, \ldots, n$. It is evident that $p_{k,r} = 0$, if $r < k$ or if $r > N - n + k$. Now we will consider the case, when $k \leqslant r \leqslant N - n + k$. The event $\{X_{k,n,n} = x_r\}$ assumes that order statistics

$$X_{1,n,N}, X_{2,n,N}, \ldots, X_{k-1,n,N}$$

take on $(k-1)$ arbitrary distinct values, which are less than x_r, while order statistics

$$X_{k+1,n,N} < X_{k+2,n,N} < \cdots < X_{n,n,N}$$

take on $(n-k)$ arbitrary values, which are greater than x_r. The combinatorial arguments show that there are

$$\frac{(r-1)(r-2)\cdots(r-k+1)}{(k-1)!}$$

options to get appropriate values for

$$X_{1,n,N}, X_{2,n,N}, \ldots, X_{k-1,n,N}$$

and

$$\frac{(N-r)(N-r-1)\cdots(N-n-r+k+1)}{(n-k)!}$$

variants for

$$X_{k+1,n,N}, X_{k+2,n,N}, \ldots, X_{n,n,N}.$$

Then it follows from (6.2) that

$$p_{k,r} = \frac{(r-1)(r-2)\cdots(r-k+1)(N-r)(N-r-1)\cdots(N-n-r+k+1)n!}{N(N-1)\cdots(N-n+1)(k-1)!(n-k)!}$$

$$= \binom{r-1}{k-1}\binom{N-r}{n-k} \Big/ \binom{N}{n}, \qquad (6.3)$$

for any $1 \leqslant k \leqslant n$ and $k \leqslant r \leqslant N - n + k$.

Exercise 6.14. Find the joint distribution of order statistics $X_{i,n,n}$ and $X_{j,n,n}$ for $1 \leqslant i < j \leqslant n = N$.

Exercise 6.15. Find the joint distribution of order statistics

$$X_{1,n,N}, \; X_{2,n,N}, \ldots, \; X_{k,n,N}.$$

Some interesting results can be obtained for conditional distributions of order statistics $X_{i,n,N}$.

Example 6.2. Consider the conditional probabilities

$$P\{X_{1,n,N} = x_{r(1)}, X_{2,n,N} = x_{r(2)}, \ldots, X_{k-1,n,N} = x_{r(k-1)} \mid X_{k,n,n} = x_{r(k)}\}.$$

Indeed, these probabilities are defined if $k \leqslant r(k) \leqslant N - n + k$, when

$$P\{X_{r,n,N} = x_{k(r)}\} > 0,$$

and they are positive, if $1 \leqslant r(1) < r(2) < \cdots < r(k-1) < r(k)$. From example 6.1 we know that

$$P\{X_{k,n,N} = x_{r(k)}\} = \frac{\dbinom{r(k)-1}{k-1} \dbinom{N-r(k)}{n-k}}{\dbinom{N}{n}}, \qquad (6.4)$$

and it follows from exercise 6.15 that

$$P\{X_{1,n,N} = x_{r(1)}, X_{2,n,N} = x_{r(2)}, \ldots, X_{k-1,n,N} = x_{r(k-1)}, X_{k,n,N} = x_{r(k)}\} = \frac{\dbinom{N-r(k)}{n-k}}{\dbinom{N}{n}}.$$

Hence,

$$P\{X_{1,n,N} = x_{r(1)}, X_{2,n,N} = x_{r(2)}, \ldots, X_{k-1,n,N} = x_{r(k-1)} \mid X_{k,n,N} = x_{r(k)}\}$$

$$= \frac{P\{X_{1,n,N} = x_{r(1)}, X_{2,n,N} = x_{r(2)}, \ldots, X_{k-1,n,N} = x_{r(k-1)}, X_{k,n,N} = x_{r(k)}\}}{P\{X_{r,n,N} = x_{r(k)}\}} = \frac{1}{\dbinom{r(k)-1}{k-1}}.$$

Comparing with (6.2), we see that the conditional distribution of order statistics

$$X_{1,n,N}, \; X_{2,n,N}, \ldots, \; X_{k-1,n,N},$$

given that

$$X_{k,n,N} = x_{r(k)},$$

coincides with the unconditional distribution of order statistics

$$X_{1,k-1,r-1}, \ldots, X_{k-1,k-1,r-1},$$

which correspond to sampling without replacement from the set of population values $\{x_1, x_2, \ldots, x_{r-1}\}$.

The similar arguments show that the conditional distribution of

$$X_{k+1,n,N}, X_{k+2,n,N}, \ldots, X_{n,n,N},$$

given that

$$X_{k,n,N} = r,$$

coincides with the unconditional distribution of order statistics

$$Y_{1,n-k,N-r}, \ldots, Y_{n-k,n-k,N-r}.$$

which correspond to sampling without replacement from the set $\{x_{r-1}, \ldots, x_n\}$.

Moreover, it can be shown that vectors

$$(X_{1,n,N}, X_{2,n,N}, \ldots, X_{k-1,n,N})$$

and

$$(X_{k+1,n,N}, X_{k+2,n,N}, \ldots, X_{n,n,N})$$

are conditionally independent given that $X_{k,n,N}$ is fixed. The simplest case of this statement is considered in the next exercise.

Exercise 6.16. Show that for any $2 \leqslant r \leqslant N-1$ order statistics $X_{1,3,N}$ and $X_{3,3,N}$ are conditionally independent given that $X_{2,3,N} = x_r$.

Example 6.3. Consider the binomial population with

$$P(X = x) = \binom{n}{x} p^x (1-p)^{n-x}, \quad x = 0, 1, \ldots, n, \ n \geqslant 1,$$

and

$$F(x) = P(X \leqslant x) = \sum_{j=0}^{x} \binom{n}{x} p^j (1-p)^{n-j}.$$

For the r-th order statistic

$$P(X_{r,n} \leqslant x) = \sum_{j=r}^{n} \binom{n}{j} (F(x))^j (1-F(x))^{n-j}, \quad x = 0, 1, \ldots, n$$

The above expression allows us the following equality

$$P(X_{r,n} = x) = I_{F(x)}(r, n-r+1) - I_{F(x-1)}(r, n-r+1);$$

where $I_\alpha(a,b)$ is the incomplete Beta function defined as

$$I_\alpha(a,b) = \frac{1}{B(a,b)} \int_0^\alpha x^{a-1}(1-x)^{b-1} dx.$$

Exercise 6.17. Consider the Poisson distribution with

$$P(X = x) = \frac{\lambda^x}{x!} e^{-x}, \quad x = 0, 1, 2, \ldots, \quad \lambda > 0.$$

Show that

$$P(X_{r,n} = x) = I_{F(x)}(r, n - r + 1) - I_{F(x-1)}(r, n - r + 1),$$

where

$$F(x) = P(X \leqslant x) = e^{-\lambda} \sum_{j=0}^{x} \frac{\lambda^j}{j!}$$

and the incomplete Beta function $I\,w(a, b)$ is defined above.

Check your solutions

Exercise 6.1 (solution). In this case the sample size $n = 2$ and

$$P\{X_{1,2} = X_{2,2}\} = P\{X_1 = X_2\} = \sum_r P\{X_1 = x_r, X_2 = x_r\}$$

$$= \sum_r P\{X_1 = x_r\} P\{X_2 = x_r\} = \sum_r p_r^2. \tag{6.5}$$

It is evident now that

$$0 \leqslant p^2 \leqslant \sum_r p_r^2 \leqslant p \sum_r p_r = p.$$

Exercise 6.2 (solution). In the general case it follows from (6.5) that

$$\pi_2 = 1 - \sum_r p_r^2.$$

Then, in our partial case

$$\pi_2 = 1 - 1/n.$$

Since

$$P\{X_{\alpha(1)} = 1, X_{\alpha(2)} = 2, \ldots, X_{\alpha(n)} = n\} = P\{X_{\alpha(1)} = 1\} P\{X_{\alpha(2)} = 2\} \cdots P\{X_{\alpha(n)} = n\}$$

$$= P\{X = 1\} P\{X_2 = 2\} \cdots P\{X_n = n\} = \frac{1}{n^n},$$

for any of $n!$ permutations $(\alpha(1), \alpha(2), \ldots, \alpha(n))$ of numbers $(1, 2, \ldots, n)$, one evidently obtains that

$$\pi_n = n! P\{X_1 = 1, X_2 = 2, \ldots, X_n = n\} = \frac{n!}{n^n}.$$

One obtains analogously that

$$\pi_r = \frac{n(n-1) \cdots (n-r+1)}{n^r}, \quad \text{if } 2 < r < n.$$

Exercise 6.3 (solution). Evidently,

$$P\{X_{k,n} = 0\} = P\{N(n) \geqslant k\},$$

here $N(n)$ denotes the number of X's among $x_1, x_2, \ldots, x_n$, which are zero. We have that

$$P\{N(n) = m\} = \binom{n}{m} p^m g^{n-m}$$

and

$$P\{X_{k,n} = 0\} = \sum_{m=k}^{n} \binom{n}{m} p^m g^{n-m}.$$

Hence,

$$P\{X_{k,n} = 1\} = 1 - P\{X_{k,n} = 0\} = 1 - \sum_{m=k}^{n} \binom{n}{m} p^m g^{n-m} = \sum_{m=0}^{k-1} \binom{n}{m} p^m g^{n-m}.$$

Exercise 6.4 (answer).

$$P\{X_{j,n} = 0, X_{k,n} = 0\} = P\{X_{k,n} = 0\} = \sum_{m=k}^{n} \binom{n}{m} p^m g^{n-m},$$

$$P\{X_{j,n} = 0, X_{k,n} = 1\} = \sum_{m=j}^{k-1} \binom{n}{m} p^m g^{n-m}$$

and

$$P\{X_{j,n} = 1, X_{k,,n} = 1\} = P\{X_{j,n} = 1\} = \sum_{m=0}^{j-1} \binom{n}{m} p^m g^{n-m}.$$

Then,

$$P\{W_{j,k,n} = 1\} = 1 - P\{W_{j,k,n} = 0\} = P\{X_{j,n} = 0, X_{k,n} = 1\} = \sum_{m=j}^{k-1} \binom{n}{m} p^m g^{n-m}.$$

Exercise 6.5 (solution). We note that $P\{X \geqslant k\} = p^k$ and hence

$$P\{X \leqslant k\} = 1 - P\{X \geqslant k+1\} = 1 - p^{k+1}, \quad k = 0, 1, 2, \ldots.$$

Then

$$P\{X_{1,n} \geqslant k\} = P\{X_1 \geqslant k, X_2 \geqslant k, \ldots, X_n \geqslant k\}$$

$$= P\{X_1 \geqslant k\} P\{X_2 \geqslant k\} \cdots P\{X_n \geqslant k\} = p^{kn}$$

and

$$P\{X_{1,n} = k\} = P\{X_{1,n} \geqslant k\} - P\{X_{1,n} \geqslant k+1\}$$

$$= p^{kn} - p^{(k+1)n} = (1 - p^k) p^{kn}, \quad k = 0, 1, 2, \ldots.$$

It means that $X_{1,n}$ has the geometric distribution also.

Analogously we obtain that

$$P\{X_{n,n} \leqslant k\} = P\{X_1 \leqslant k, \ldots, X_n \leqslant k\} = P^n\{X \leqslant k\} = (1-p^{k+1})^n$$

and

$$P\{X_{n,n} = k\} = P\{X_{n,n} \leqslant k\} - P\{X_{n,n} \leqslant k-1\}$$
$$= (1-p^{k+1})^n - (1-p^k)^n, \quad k = 0,1,2,\ldots.$$

Exercise 6.6 (answer).

$$P\{X_{2,n} \geqslant k\} = p^{k(n-1)}(n - (n-1)p^k)$$

and

$$P\{X_{2,n} = k\} = np^{k(n-1)}(1-p^{(n-1)}) - (n-1)p^{kn}(1-p^n), \quad k = 0,1,2,\ldots.$$

Exercise 6.7 (answer). In this case

$$P\{Y = k\} = (1-p)p^k, \quad k = 0,1,2,\ldots,$$

where $p = p_1 p_2 \cdots p_n$.

Exercise 6.8 (solution). Denote

$$F(r) = p_0 + p_1 + \cdots + p_r.$$

Then

$$P\{X_{k,n} \leqslant r\} = \sum_{m=k}^{n-1} P\{X_{m,n} \leqslant r, X_{m+1,n} > r\} + P\{X_{n,n} \leqslant r\}$$
$$= \sum_{m=k}^{n} \binom{n}{m} (F(r))^m (1-F(r))^{n-m}.$$

It was proved in exercise 2.5 that the following identity (2.6) is true:

$$\sum_{m=k}^{n} \binom{n}{m} x^m (1-x)^{n-m} = I_x(k, n-k+1),$$

where

$$I_x(a,b) = \frac{1}{B(a,b)} \int_0^x t^{a-1}(1-t)^{b-1} dt$$

is the incomplete Beta function with parameters a and b, $B(a,b)$ being the classical Beta function. Hence

$$P\{X_{k,n} \leqslant r\} = I_{F(r)}(k, n-k+1)$$

and

$$P\{X_{k,n} = r\} = P\{X_{k,n} \leqslant r\} - P\{X_{k,n} \leqslant r-1\}$$
$$= I_{F(r)}(k, n-k+1) - I_{F(r-1)}(k, n-k+1). \tag{6.6}$$

Exercise 6.9 (solution). It is evident that

$$P\{X_{1,n} \geqslant i, X_{n,n} < j\} = (P\{i \leqslant X < j\})^n = (P\{X \geqslant i\} - P\{X \geqslant j\})^n = (p^i - p^j)^n,$$

for any $0 \leqslant i < j$ and

$$P\{X_{1,n} \geqslant i, X_{n,n} < j\} = 0,$$

if $i = j$. Hence,

$$P\{X_{1,n} = i, X_{n,n=j}\} = P\{X_{1,n} \geqslant i, X_{n,n} < j+1\} - P\{X_{1,n} \geqslant i, X_{n,n} < j\}-$$

$$P\{X_{1,n} \geqslant i+1, X_{n,n} < j+1\} + P\{X_{1,n} \geqslant i+1, X_{n,n} < j\} =$$

$$(p^i - p^{j+1})^n - (p^i - p^j)^n - (p^{i+1} - p^{j+1})^n + (p^{i+1} - p^j)^n =$$

$$p^{in}((1 - p^{j-i+1})^n - (1 - p^{j-i})^n - (p - p^{j-i+1})^n + (p - p^{j-i})^n),$$

if $i \geqslant j$,

$$P\{X_{1,n} = i, X_{n,n} = i\} = (1 - p)^n p^{in}, \quad i = 0, 1, 2, \ldots.$$

and

$$P\{X_{1,n} = i, X_{n,n} = j\} = 0,$$

if $j < i$.

Exercise 6.10 (solution). One sees that

$$P\{W_n = r\} = \sum_{i=0}^{\infty} P\{X_{1,n} = i, X_{n,n} = i+r\}.$$

On applying the result of exercise 6.9, we get that

$$P\{W_n = r\} = \sum_{i=0}^{\infty} p^{in}((1 - p^{r+1})^n - (1 - p^r)^n - (p - p^{r+1})^n + (p - p^r)^n) =$$

$$\frac{(1 - p^{r+1})^n - (1 - p^r)^n - (p - p^{r+1})^n + (p - p^r)^n}{1 - p^n}, \quad r = 1, 2, \ldots,$$

and

$$P\{W_n = 0\} = \sum_{i=0}^{\infty} P\{X_{1,n} = i, X_{n,n} = i\} = \frac{(1 - p)^n}{1 - p^n}.$$

Exercise 6.11 (hint). Use the evident equality

$$P\{X_{1,n} = i, W_n = r\} = P\{X_{1,n} = i, X_{n,n} = i + r\}.$$

separately for $r = 0$ and $r = 1$. Then the joint distributions of $X_{1,n}$ and $X_{n,n}$ obtained in exercise 6.9, distributions of $X_{1,n}$ and W_n given in exercises 6.5 and 6.10 correspondingly, help you to prove that

$$P\{X_{1,n} = i, W_n = r\} = P\{X_{1,n} = i\}P\{W_n = r\}$$

for any $n = 2$, $i = 0, 1, 2, \ldots$ and $r = 0, 1, 2, \ldots$.

Exercise 6.12 (solution). For the sake of simplicity we can suppose that $X_k = k$. Let $r < k < s$. Then

$$
\begin{aligned}
P\{X_{1,3} = r, X_{3,3} = s \mid X_{2,3} = k\} &= \frac{P\{X_{1,3} = r, X_{2,3} = k, X_{3,3} = s\}}{P\{X_{2,3} = k\}} \\
&= \frac{6P\{X_1 = r, X_2 = k, X_3 = s\}}{P\{X_{2,3} = k\}} = 6p_r p_s g_k,
\end{aligned} \tag{6.7}
$$

where

$$
\begin{aligned}
g_k &= \frac{p_k}{P\{X_{2,3} = k\}} = \frac{p_k}{p_k^3 + 3F(k-1)p_k^2 + 3(1 - F(k))p_k^2 + 6F(k-1)(1 - F(k))} \\
&= \frac{1}{p_k^2 + 3F(k-1)p_k + 3(1 - F(k))p_k + 6F(k-1)(1 - F(k))/p_k}
\end{aligned} \tag{6.8}
$$

and

$$F(k) = p_1 + \cdots + p_k.$$

If $r = k < s$, then

$$P\{X_{1,3} = k, X_{3,3} = s \mid X_{2,3} = k\} = \frac{3P\{X_1 = X_2 = k, X_3 = s\}}{P\{X_{2,3} = k\}} = 3p_k p_s g_k, \tag{6.9}$$

where g_k is defined in (6.8).
If $r < k = s$, then analogously

$$P\{X_{1,3} = r, X_{3,3} = k \mid X_{2,3} = k\} = 3p_r p_s g_k, \tag{6.10}$$

and for $r = k = s$ one obtains that

$$P\{X_{1,3} = k, X_{3,3} = k \mid X_{2,3} = k\} = \frac{P\{X_1 = X_2 = X_3 = k\}}{P\{X_{2,3} = k\}} = p_k^2 g_k. \tag{6.11}$$

Now if $r < k$ we have that

$$P\{X_{1,3} = r \mid X_{2,3} = k\} = \frac{P\{X_{1,3} = r, X_{2,3} = k\}}{P\{X_{2,3} = k\}}$$

$$= \frac{P\{X_{1,3} = r, X_{2,3} = k, X_{3,3} = k\} + P\{X_{1,3} = r, X_{2,3} = k, X_{3,3} > k\}}{P\{X_{2,3} = k\}}$$

$$= \frac{3p_r p_k^2 + 6p_r p_k(1 - F(k))}{P\{X_{2,3} = k\}} = 3p_r(p_k + 2 - 2F(k))g_k. \tag{6.12}$$

If $r = k$, then

$$P\{X_{1,3} = k \mid X_{2,3} = k\} = \frac{P\{X_{1,3} = X_{2,3} = X_{3,3} = k\} + P\{X_{1,3} = X_{2,3} = k, X_{3,3} > k\}}{P\{X_{2,3} = k\}}$$

$$= (p_k^2 + 3p_k(1 - F(k)))g_k. \tag{6.13}$$

Analogously,

$$P\{X_{3,3} = s \mid X_{2,3} = k\} = 3p_s(p_k + 2F(k-1))g_k, \tag{6.14}$$

if $s > k$, and

$$P\{X_{3,3} = k \mid X_{2,3} = k\} = (p_k^2 + 3p_k F(k-1))g_k. \tag{6.15}$$

Exercise 6.13 (solution). In fact, to solve this problem we need to check that the following four equalities are valid:

$$P\{X_{1,3} = a, X_{3,3} = b \mid X_{2,3} = a\} = P\{X_{1,3} = a \mid X_{2,3} = a\}P\{X_{3,3} = b \mid X_{2,3} = a\},$$

$$P\{X_{1,3} = a, X_{3,3} = b \mid X_{2,3} = b\} = P\{X_{1,3} = a \mid X_{2,3} = b\}P\{X_{3,3} = b \mid X_{2,3} = b\},$$

$$P\{X_{1,3} = a, X_{3,3} = a \mid X_{2,3} = a\} = P\{X_{1,3} = a \mid X_{2,3} = a\}P\{X_{3,3} = a \mid X_{2,3} = a\}$$

and

$$P\{X_{1,3} = b, X_{3,3} = b \mid X_{2,3} = b\} = P\{X_{1,3} = b \mid X_{2,3} = b\}P\{X_{3,3} = b \mid X_{2,3} = b\}.$$

It is evident that

$$P\{X_{1,3} = a \mid X_{2,3} = a\} = 1, \quad P\{X_{3,3} = b \mid X_{2,3} = b\} = 1,$$

$$P\{X_{1,3} = a, X_{3,3} = b \mid X_{2,3} = b\} = P\{X_{1,3} = a \mid X_{2,3} = b\},$$

$$P\{X_{1,3} = a, X_{3,3} = b \mid X_{2,3} = a\} = P\{X_{3,3} = b \mid X_{2,3} = a\}$$

and these relations immediately imply that

$$P\{X_{1,3} = a, X_{3,3} = b \mid X_{2,3} = b\} = P\{X_{1,3} = a \mid X_{2,3} = b\}P\{X_{3,3} = b \mid X_{2,3} = b\}$$

and

$$P\{X_{1,3} = a, X_{3,3} = b \mid X_{2,3} = a\} = P\{X_{1,3} = a \mid X_{2,3} = a\}P\{X_{3,3} = b \mid X_{2,3} = a\}.$$

Then, we see that

$$P\{X_{1,3} = a, X_{3,3} = a \mid X_{2,3} = a\} = 1 - P\{X_{1,3} = a, X_{3,3} = b \mid X_{2,3} = a\}$$

$$= P\{X_{3,3} = b \mid X_{2,3} = a\} = P\{X_{3,3} = a \mid X_{2,3} = a\}$$

and hence

$$P\{X_{1,3} = a, X_{3,3} = a \mid X_{2,3} = a\} = P\{X_{1,3} = a \mid X_{2,3} = a\}P\{X_{3,3} = a \mid X_{2,3} = a\}.$$

Analogously one obtains that

$$P\{X_{1,3} = b, X_{3,3} = b \mid X_{2,3} = b\} = P\{X_{1,3} = b \mid X_{2,3} = b\}P\{X_{3,3} = b \mid X_{2,3} = b\}.$$

Exercise 6.14 (answer).

$$P\{X_{i,n,N} = x_r, X_{j,n,N} = x_s\} = \begin{cases} \dfrac{\dbinom{r-1}{i-1}\dbinom{s-r-1}{j-i-1}\dbinom{N-s}{n-j}}{\dbinom{N}{n}}, & \text{if } i \leqslant r < s \leqslant N-n+j \\ & \text{and } s-r \geqslant j-i, \text{and} \\ 0, & \text{otherwise.} \end{cases}$$

$$(6.16)$$

Exercise 6.15 (answer).

$$P\{X_{1,n,N} = x_{r(1)}, X_{2,n,N} = x_{r(2)}, \ldots, X_{k,n,N} = x_{r(k)}\} = \frac{\dbinom{N-r(k)}{n-k}}{\dbinom{N}{n}}$$

if $1 \leqslant r(1) < r(2) < \cdots < r(k)$ and $N - r(k) \geqslant n - k$, while

$$P\{X_{1,n,N} = x_{r(1)}, X_{2,n,N} = x_{r(2)}, \ldots, X_{k,n,N} = x_{r(k)}\} = 0,$$

otherwise.

Exercise 6.16 (solution). One needs to show that for any $1 \leqslant i < r < j \leqslant N$ the following relation holds:

$$P\{X_{1,3,N} = x_i, X_{3,3,N} = x_j \mid X_{2,3,N} = x_r\}$$

$$= P\{X_{1,3,N} = x_i \mid X_{2,3,N} = x_r\} P\{X_{3,3,N} = x_j \mid X_{2,3,N} = x_r\}.$$

This is equivalent to the equality

$$P\{X_{1,3,N} = x_i, X_{2,3,N} = x_r, X_{3,3,N} = x_j\} P\{X_{2,3,N} = x_r\}$$

$$= P\{X_{1,3,N} = x_i, X_{2,3,N} = x_r\} P\{X_{2,3,N} = x_r, X_{3,3,n} = x_j\}. \tag{6.17}$$

To check this equality we must recall from (6.2), (6.3) and (6.16) that

$$P\{X_{1,3,N} = x_i, X_{2,3,N} = x_r, X_{3,3,N} = x_j\} = \frac{1}{\binom{N}{3}},$$

$$P\{X_{1,3,N} = x_i, X_{2,3,N} = x_r\} = \frac{N-r}{\binom{N}{3}},$$

$$P\{X_{2,3,N} = x_r, X_{3,3,N} = x_j\} = \frac{r-1}{\binom{N}{3}}$$

and

$$P\{X_{2,3,n} = x_r\} = \frac{(r-1)(N-r)}{\binom{N}{3}}.$$

The rest part of the proof is evident.

Exercise 6.17 (Hint). See Example 6.3.

Chapter 7

Moments of Order Statistics: General Relations

The main relations are given for moments of order statistics.

Taking into account equality (2.8) we get the general formula for moments of order statistics $X_{k,n}$ related to a population with a d.f. F. In fact,

$$\mu_{k:n}^{(r)} = E(X_{k,n})^r = \int_{-\infty}^{\infty} x^r dF_{k:n}(x)$$

$$= \frac{n!}{(k-1)!(n-k)!} \int_{-\infty}^{\infty} x^r (F(x))^{k-1}(1-F(x))^{n-k} dF(x). \tag{7.1}$$

For continuous distribution functions F this equality can be expressed as

$$\mu_{k:n}^{(r)} = \frac{n!}{(k-1)!(n-k)!} \int_0^1 (G(u))^r u^{k-1}(1-u)^{n-k} du, \tag{7.2}$$

where $G(u)$ is the inverse of F. For absolutely continuous distributions with probability density function f the RHS of (7.1) coincides with

$$\frac{n!}{(k-1)!(n-k)!} \int_{-\infty}^{\infty} x^r (F(x))^{k-1}(1-F(x))^{n-k} f(x) dx. \tag{7.3}$$

Similar relations are valid for joint (product) moments of order statistics. For the sake of simplicity we consider joint moments

$$\mu_{k(1),k(2):n}^{(r(1),r(2))} = E(X_{k(1),n})^{r(1)}(X_{k(2),n})^{r(2)}, \quad 1 \leqslant k(1) < k(2) \leqslant n,$$

of two order statistics only. From (2.14) we obtain for absolutely continuous distributions that

$$\mu_{k(1),k(2):n}^{(r(1),r(2))} = c(k(1),k(2),n) \int_{-\infty}^{\infty} \int_{-\infty}^{x} x^{r(1)} y^{r(2)} (F(x))^{k(1)-1}$$

$$\times (F(y)-F(x))^{k(2)-k(1)-1}(1-F(y))^{n-k(2)} f(x)f(y) dx dy, \tag{7.4}$$

where

$$c(k(1),k(2),n) = \frac{n!}{(k(1)-1)!(k(2)-k(1)-1)!(n-k(2))!}. \tag{7.5}$$

M. Ahsanullah et al., *An Introduction to Order Statistics*,
Atlantis Studies in Probability and Statistics 3,
DOI: 10.2991/978-94-91216-83-1_7, © Atlantis Press 2013

In the general case

$$\mu_{k(1),k(2):n}^{(r(1),r(2))} = c(k(1),k(2),n) \int_{-\infty}^{\infty} \int_{-\infty}^{x} X^{r(1)} y^{r(2)} (F(x))^{k(1)-1}$$

$$\times (F(y) - F(x))^{k(2)-k(1)-1} (1 - F(y))^{n-k(2)} dF(x) dF(y), \qquad (7.6)$$

where $c(r(1), r(2), n)$ is given in (7.5). We will use the following notations also:

$$\mu_{k:n} = EX_{k,n}$$

will be applied for the sake of simplicity instead of $\mu_{k:n}^{(1)}$.

$$\mu_{k(1)k(2):n} = E\left(X_{k(1),n} X_{k(2),n}\right)$$

will change $\mu_{k(1),k(2):n}^{(1,1)}$;

$$\text{Var}(X_{k,n}) = \mu_{k:n}^{(2)} - (\mu_{k:n})^2$$

will denote the variance of $X_{k,n}$;

$$\text{cov}(X_{r,n}, X_{s,n}) = \mu_{r,s:n} - \mu_{r:n}\mu_{s:n}$$

will be used for the covariance between $X_{r,n}$ and $X_{s,n}$.

It is interesting to find the corresponding conditions, which provide the existence of different moments of order statistics.

Example 7.1. Let there exist the population moment $\alpha_r = E|X|^r$, i.e.,

$$E|X|^r = \int_{-\infty}^{\infty} |x|^r dF(x) < \infty. \qquad (7.7)$$

Then due to (7.7) we easily derive that

$$E|X_{k,n}|^r \leqslant \frac{n!}{(k-1)!(n-k)!} \int_{-\infty}^{\infty} |x|^r (F(x))^{k-1} (1 - F(x))^{n-k} dF(x)$$

$$\leqslant \frac{n!}{(k-1)!(n-k)!} \int_{-\infty}^{\infty} |x|^r dF(x)$$

$$= \frac{n!}{(k-1)!(n-k)!} E|X|^r < \infty. \qquad (7.8)$$

It follows from (7.8) that the existence of the moment α_r implies the existence of all moments

$$E|X_{k,n}|^r, \quad 1 \leqslant k \leqslant n, \ n = 1, 2, \dots.$$

Exercise 7.1. Show that if

$$E|X|^r = \infty$$

for some r, then for any $n = 1, 2, \dots$ it is possible to find such order statistic $X_{k,n}$ that

$$E|X_{k,n}|^r = \infty.$$

Remark 7.1. Since

$$P\{X_{1,n} \leqslant X_{k,n} \leqslant X_{n,n}\} = 1$$

for any $1 < k < n$, one has the evident inequality

$$E|X_{k,n}|^r \leqslant E(|X_{1,n}|^r + |X_{n,n}|^r).$$

Hence, if $E|X_{k,n}|^r = \infty$, then at least one of equalities $E|X_{1,n}|^r = \infty$ or $E|X_{n,n}|^r = \infty$ is valid.

Exercise 7.2. Let $E|X_{k,n}| = \infty$ and $E|X_{k+1,n}| < \infty$. Show that then $E|X_{r,n}| = \infty$, if $r = 1, 2, \ldots, k-1$. Analogously, if $E|X_{k,n}| = \infty$ and $E|X_{k-1,n}| < \infty$, then $E|X_{r,n}| = \infty$, for $r = k+1, \ldots, n$.

Exercise 7.3. Let X have the Cauchy distribution with the density function

$$f(x) = \frac{1}{\pi(1+x^2)}.$$

Show that for any $r = 1, 2, \ldots$, relation

$$E|X_{k,n}|^r < \infty$$

holds if $r < k < n - r + 1$.

Example 7.2. Consider a more general situation than one given in exercise 7.3. The following result was proved by Sen (1959).

Let $E|X|^\alpha < \infty$. We will show that then moments $\mu_{k:n}^{(r)}$ exist for all k such that $r/\alpha \leqslant k \leqslant (n-r+1)/\alpha$.

Due to the result given in example 7.1, this statement is evident if $r/\alpha \leqslant 1$, since the existence of the moment $E|X|^\alpha$ implies the existence of moments $\mu_{k:n}^{(r)}$, $k = 1, 2, \ldots, n$, $n = 1, 2, \ldots$, for $r \leqslant \alpha$. Hence, we need to consider the case $r > \alpha$ only.

If $E|X|^\alpha < \infty$, then integrals

$$I_1(\alpha) = \int_0^\infty x^{\alpha-1} F(-x)dx$$

and

$$I_2(\alpha) = \int_0^\infty x^{\alpha-1}(1 - F(x))dx$$

are finite and

$$E|X|^\alpha = \alpha(I_1(\alpha) + I_2(\alpha)). \tag{7.9}$$

Moreover, if it is known that $I_1(\alpha)$ and $I_2(\alpha)$ are finite, then (7.9) is also true.

Note, that if $E|X|^{\alpha} < \infty$ (or simply, if both integrals, $I_1(\alpha)$ and $I_2(\alpha)$ are finite) then

$$(1 - F(x)) = o(x^{-\alpha}), \quad F(-x) = o(x^{-\alpha}), \quad x \to \infty. \tag{7.10}$$

It suffices for us to prove that if

$$E|X|^{\alpha} < \infty \text{ and } \frac{r}{\alpha} \leqslant k \leqslant \frac{n-r+1}{\alpha},$$

then

$$\int_0^{\infty} x^{r-1} F_{k,n}(-x) dx < \infty, \quad \int_0^{\infty} x^{r-1}(1 - F_{k,n}(x)) dx < \infty,$$

and thus,

$$E|X_{k,n}|^r < \infty.$$

Let us recall (see relation (2.5)) that

$$F_{k,n}(x) = \sum_{m=k}^{n} \binom{n}{m} (F(x))^m (1 - F(x))^{n-m} \tag{7.11}$$

and

$$1 - F_{k,n}(x) = \sum_{m=0}^{k-1} \binom{n}{m} (F(x))^m (1 - F(x))^{n-m}. \tag{7.12}$$

The following evident inequalities are valid for the LHS of (7.12):

$$1 - F_{k,n}(x) \leqslant (1 - F(x))^{n-k+1} \sum_{m=0}^{k-1} \binom{n}{m} (F(x))^m (1 - F(x))^{k-m-1}$$

$$\leqslant (1 - F(x))^{n-k+1} \sum_{m=0}^{k-1} \binom{n}{m}$$

$$\leqslant (1 - F(x))^{n-k+1} \sum_{m=0}^{n} \binom{n}{m}$$

$$= 2^n (1 - F(x))^{n-k+1}.$$

Further,

$$0 \leqslant \int_0^{\infty} x^{r-1}(1 - F_{k,n}(x)) dx$$

$$\leqslant 2^n \int_0^{\infty} x^{r-1}(1 - F(x))^{n-k+1}$$

$$= 2^n \int_0^{\infty} x^{\alpha-1}(1 - F(x)) h_k(x) dx, \tag{7.13}$$

where

$$h_k(x) = x^{r-\alpha}(1 - F(x))^{n-k},$$

and due to (7.10),

$$h_k(x) = o(x^{r-\alpha-(n-k)\alpha}) = o(1), \quad x \to \infty, \tag{7.14}$$

if $k \leqslant (n-r+1)/\alpha$.

Since

$$\int_0^\infty x^{\alpha-1}(1-F(x))dx < \infty,$$

it follows evidently from (7.13) and (7.14) that

$$\int_0^\infty x^{r-1}(1-F_{k,n}(x))dx < \infty. \tag{7.15}$$

Similarly, one gets that

$$\int_0^\infty x^{r-1}F_{k,n}(-x)dx < \infty, \tag{7.16}$$

if $k \geqslant r/\alpha$. Finally, (7.12) and (7.13) imply that

$$E|X_{k,n}|^r < \infty,$$

if $r/\alpha \leqslant k \leqslant (n-r+1)/\alpha$.

Remark 7.2. All saying in example 7.2 enables us to state additionally that if $X \geqslant 0$ and $EX^\alpha < \infty$, then

$$0 \leqslant EX_{k,n}^r < \infty$$

for all k and n such that

$$1 \leqslant k \leqslant \frac{n-r+1}{\alpha}.$$

If $X \leqslant 0$ and $E(-X)^\alpha < \infty$, then

$$0 \leqslant E(-X_{k,n})^r < \infty, \quad \frac{r}{\alpha} \leqslant k \leqslant n.$$

Some useful relations for moments come from the evident identity

$$X_{1,n} + \cdots + X_{n,n} = X_1 + \cdots + X_n \tag{7.17}$$

and related equalities. For instance, the simplest corollary of (7.17) is as follows:

$$E(X_{1,n} + \cdots + X_{n,n}) = E(X_1 + \cdots + X_n) = nEX, \tag{7.18}$$

if there exists a population expectation. Say, if $EX = 0$ and $n = 2$ in (7.18), then we get that

$$EX_{2,2} = -EX_{1,2}.$$

A natural generalization of (7.17) has the form

$$g\left(\sum_{k=1}^n h(X_{k,n})\right) = g\left(\sum_{k=1}^n h(X_k)\right), \tag{7.19}$$

where $g(x)$ and $h(x)$ are arbitrary functions.

Example 7.3. The following identities based on (7.19) can be useful in some situations:

$$E\left(\sum_{k=1}^{n} X_{k,n}^m\right)^r = E\left(\sum_{k=1}^{n} X_k^m\right)^r, \quad m = 1, 2, \ldots, \quad r = 1, 2, \ldots. \tag{7.20}$$

The case $m = r = 1$ was considered in (7.18). Similarly we get that

$$\sum_{k=1}^{n} EX_{k,n}^m = \sum_{k=1}^{n} EX_k^m = nEX^m \tag{7.21}$$

for any m provided that the corresponding moment EX^m exists.

If now $m = 1$ and $r = 2$ in (7.20), one gets that

$$\sum_{k=1}^{n} EX_{k,n}^2 + 2\sum_{k=1}^{n-1}\sum_{r=k+1}^{n} EX_{k,n}X_{r,n} = E\left(\sum_{k=1}^{n} X_k\right)^2 = nEX^2 + n(n-1)(EX)^2. \tag{7.22}$$

Due to equality (7.21) (for $m = 2$), identity (7.22) can be simplified as follows:

$$\sum_{k=1}^{n-1}\sum_{r=k+1}^{n} EX_{k,n}X_{r,n} = \frac{n(n-1)(EX)^2}{2}. \tag{7.23}$$

Exercise 7.4. Prove that the following identity holds for covariances between order statistics:

$$\sum_{k=1}^{n}\sum_{r=1}^{n} \text{Cov}\left(EX_{k,n}, X_{r,n}\right) = n\,\text{Var}\,X. \tag{7.24}$$

Example 7.4. One more similar identity

$$\sum_{k=1}^{n}\sum_{r=1}^{n} X_{k,n}^m X_{r,n}^s = \sum_{k=1}^{n}\sum_{r=1}^{n} X_k^m X_r^s \tag{7.25}$$

implies the relation

$$\sum_{k=1}^{n}\sum_{r=1}^{n} EX_{k,n}^m X_{r,n}^s = nEX^{m+s} + n(n-1)EX^m EX^s. \tag{7.26}$$

Exercise 7.5. Prove that the following relation holds for any $1 \leqslant k \leqslant n-1$ and $m = 1, 2, \ldots$:

$$kEX_{k+1,n}^m + (n-k)EX_{k,n}^m = nEX_{k,n-1}^m. \tag{7.27}$$

Check your solutions

Exercise 7.1 (solution). Suppose that all order statistics have finite moments $E|X_{k,n}|^r < \infty$. It is clear that

$$P\{X_{1,n} \leqslant X_1 \leqslant X_{n,n}\} = 1.$$

Then,

$$P\{|X_1|^r \leqslant \max\{|X_{1,n}|^r, |X_{n,n}|^r\} = 1$$

and the following inequality, which contradicts to the initial condition is valid:

$$E|X|^r = E|X_1|^r \leqslant E\max\{|X_{1,n}|^r, |X_{n,n}|^r\} \leqslant E(|X_{1,n}|^r + |X_{n,n}|^r) < \infty.$$

Hence, there is at least one order statistic such that $E|X_{k,n}|^r = \infty$.

Exercise 7.2 (hint). In the first case consider the evident inequalities

$$E|X_{k,n}| \leqslant E(|X_{r,n}| + |X_{k+1,n}|), \quad r = 1, 2, \ldots, k-1.$$

In the second case it suffices to use inequalities

$$E|X_{k,n}| \leqslant E(|X_{r,n}| + |X_{k-1,n}|), \quad r = k+1, k+2, \ldots, n.$$

Exercise 7.3 (hint). Prove that

$$G(x) \sim \frac{1}{\pi(1-x)}, \quad x \to 1,$$

$$G(x) \sim -\frac{1}{\pi x}, \quad x \to 0.$$

It means that

$$(G(u))^r u^{k-1} (1-u)^{n-k} \to \frac{1}{\pi^r} u^{k-r-1} \text{ as } u \to 0,$$

and

$$(G(u))^r u^{k-1} (1-u)^{n-k} \to \frac{1}{\pi^r} (1-u)^{n-k-r1} \text{ as } u \to 1.$$

Now it becomes evident that the integral is finite if $k - r - 1 > -1$ and $n - k - r > -1$, that is

$$E|X_{k,n}|^r < \infty$$

if and only if $r < k < n - r + 1$.

Exercise 7.4 (hint). Use the evident identity

$$\left(\sum_{k=1}^{n} (X_{k,n} - EX_{k,n}) \right)^2 = \left(\sum_{k=1}^{n} (X_k - EX_k) \right)^2.$$

Exercise 7.5 (solution). Recall that

$$E(X_{r,n})^m = \frac{n!}{(r-1)!(n-r)!} \int_{-\infty}^{\infty} x^m (F(x))^{r-1} (1 - F(x))^{n-r} dF(x).$$

Then

$$kEX_{k+1,n}^m + (n-k)EX_{k,n}^m = \frac{k(n!)}{k!(n-k-1)!} \int_{-\infty}^{\infty} x^m (F(x))^k (1 - F(x))^{n-k-1} dF(x)$$

$$+ \frac{(n-k)(n!)}{(k-1)!(n-k)!} \int_{-\infty}^{\infty} x^m (F(x))^{k-1} (1 - F(x))^{n-k} dF(x)$$

$$= \frac{n!}{(k-1)!(n-k-1)!} \int_{-\infty}^{\infty} x^m \{ (F(x))^k (1 - F(x))^{n-k-1}$$

$$+ (F(x))^{k-1} (1 - F(x))^{n-k} \} dF(x)$$

$$= \frac{n!}{(k-1)!(n-k-1)!} \int_{-\infty}^{\infty} x^m \{ (F(x))^{k-1} (1 - F(x))^{n-k-1} \} dF(x).$$

The latter expression coincides evidently with

$$nEX_{k,n-1}^m.$$

Chapter 8

Moments of Uniform and Exponential Order Statistics

In chapter 4 we proved some representations for uniform and exponential order statistics, which enable us to express these order statistics via sums or products of independent random variables. By virtue of the corresponding expressions one can easily find single and joint moments of exponential and uniform order statistics.

8.1 Uniform order statistics

Indeed, in the case of the standard uniform distribution one can use expression (7.3) to find single moments of order statistics $U_{k,n}$. In fact, in this case for any $\alpha > -k$ we get the following result:

$$
\begin{aligned}
E(U_{k,n})^\alpha &= \frac{n!}{(k-1)!(n-k)!} \int_0^1 x^\alpha x^{k-1}(1-x)^{n-k}dx \\
&= \frac{n!}{(k-1)!(n-k)!} B(\alpha+k, n-k+1) \\
&= \frac{n!\Gamma(\alpha+k)\Gamma(n-k+1)}{(k-1)!(n-k)!\Gamma(n+\alpha+1)} \\
&= \frac{n!\Gamma(\alpha+k)}{(k-1)!\Gamma(n+\alpha+1)}
\end{aligned}
\tag{8.1}
$$

where $B(a,b)$ and $\Gamma(s)$ denote the beta function and the gamma function respectively, which are tied by the relation

$$
B(a,b) = \frac{\Gamma(a)\Gamma(b)}{\Gamma(a+b)}.
$$

Note also that

$$
\Gamma(n) = (n-1)! \text{ for } n = 1, 2, \ldots.
$$

If α is an integer, then the RHS on (8.1) is simplified. For instance,

$$
EU_{k,n} = \frac{n!\Gamma(k+1)}{(k-1)!\Gamma(n+2)} = \frac{n!k!}{(k-1)!(n+1)!} = \frac{k}{n+1}, \quad 1 \leqslant k \leqslant n,
\tag{8.2}
$$

M. Ahsanullah et al., *An Introduction to Order Statistics*,
Atlantis Studies in Probability and Statistics 3,
DOI: 10.2991/978-94-91216-83-1_8, © Atlantis Press 2013

and

$$E(1/U_{k,n}) = \frac{n!\Gamma(k-1)}{(k-1)!\Gamma(n)} = \frac{n}{k-1}, \quad 2 \leqslant k \leqslant n. \tag{8.3}$$

Similarly,

$$E(U_{k,n})^2 = \frac{k(k+1)}{(n+1)(n+2)}, \quad 1 \leqslant k \leqslant n, \tag{8.4}$$

and

$$E\left(\frac{1}{(U_{k,n})^2}\right) = \frac{n(n-1)}{(k-1)(k-2)}, \quad 3 \leqslant k \leqslant n. \tag{8.5}$$

In general form, for $r = 1, 2, \ldots$, we have

$$E(U_{k,n})^r = \frac{k(k+1)\cdots(k+r-1)}{(n+1)(n+2)\cdots(n+r)}, \quad 1 \leqslant k \leqslant n, \tag{8.6}$$

and

$$E\left(\frac{1}{(U_{k,n})^r}\right) = \frac{n(n-1)\cdots(n-r+1)}{(k-1)(k-2)\cdots(k-r)}, \quad r+1 \leqslant k \leqslant n. \tag{8.7}$$

It follows from (8.2) and (8.4) that

$$\operatorname{Var}(U_{k,n}) = \frac{k(n-k+1)}{(n+1)^2(n+2)}, \quad 1 \leqslant k \leqslant n. \tag{8.8}$$

Exercise 8.1. Find the variance of $1/U_{k,n}$.

Exercise 8.2. Find the third central moments of $U_{k,n}$.

Some of the just-given moments can be obtained by means of representations (4.24) and (4.29).

Example 8.1. We know from (4.29) that

$$EU_{k,n} = E(S_k \mid S_{n+1} = 1),$$

where

$$S_n = v_1 + v_2 + \cdots + v_n, \quad n = 1, 2, \ldots,$$

and $v_1, v_2, \ldots$ are independent identically distributed random variables having the standard exponential distribution. Further, the symmetry arguments enable us to see that

$$\begin{aligned}
E(S_k \mid S_{n+1} = 1) &= \sum_{r=1}^{k} E(v_r \mid v_1 + v_2 + \cdots + v_{n+1} = 1) \\
&= kE(v_1 \mid v_1 + v_2 + \cdots + v_{n+1} = 1) \\
&= \frac{k}{n+1}(v_1 + v_2 + \cdots + v_{n+1} \mid v_1 + v_2 + \cdots + v_{n+1} = 1) \\
&= \frac{k}{n+1}.
\end{aligned}$$

From example 4.6 we know that

$$U_{k,n} \stackrel{d}{=} \frac{S_k}{S_{n+1}}$$

and $\frac{S_k}{S_{n+1}}$ is independent on the sum $S_{n+1} = v_1 + v_2 + \cdots + v_{n+1}$. Then

$$E(U_{k,n})^\alpha = E\left(\frac{S_k}{S_{n+1}}\right)^\alpha.$$

Due to the independence of $\frac{S_k}{S_{n+1}}$ and S_{n+1} we have the following relation:

$$E(S_k)^\alpha = E\left(\frac{S_k}{S_{n+1}}S_{n+1}\right)^\alpha = E\left(\frac{S_k}{S_{n+1}}\right)^\alpha E(S_{n+1})^\alpha.$$

Thus,

$$E\left(\frac{S_k}{S_{n+1}}\right)^\alpha = \frac{E(S_k)^\alpha}{E(S_{n+1})^\alpha}.$$

Now we must recall that S_m has gamma distribution with parameter m and hence

$$E(S_m)^\alpha = \frac{1}{(m-1)!}\int_0^\infty x^{\alpha+m-1}e^{-x}dx = \frac{\Gamma(\alpha+m)}{\Gamma(m)}.$$

Finally,

$$E(U_{k,n})^\alpha = E\left(\frac{S_k}{S_{n+1}}\right)^\alpha = \frac{E(S_k)^\alpha}{E(S_{n+1})^\alpha} = \frac{\Gamma(\alpha+k)\Gamma(n+1)}{\Gamma(k)\Gamma(\alpha+n+1)} = \frac{n!\,\Gamma(\alpha+k)}{(k-1)!\,\Gamma(\alpha+n+1)}$$

and the latter expression coincides with (8.1).

One more representation, (4.20), enables us to get joint (product) moments of the uniform order statistics.

Example 8.2. Consider two uniform order statistics $U_{r,n}$ and $U_{s,n}$, $1 \leqslant r < s \leqslant n$. As it was shown in (4.20),

$$(U_{1,n}, U_{2,n}, \ldots, U_{n,n}) \stackrel{d}{=} (W_1 W_2^{1/2} \cdots W_{n-1}^{1/(n-1)} W_n^{1/n}, W_2^{1/2} \cdots W_{n-1}^{1/(n-1)} W_n^{1/n}, \ldots, W_n^{1/n}),$$

where $W_1, W_2, \ldots$ are independent and have the same standard uniform distribution. Hence,

$$\begin{aligned}
EU_{r,n}U_{s,n} &= E(W_r^{1/r}W_{r+1}^{1/(r+1)} \cdots W_n^{1/n} W_s^{1/s} W_{s+1}^{1/(s+1)} \cdots W_n^{1/n}) \\
&= E(W_r^{1/r}W_{r+1}^{1/(r+1)} \cdots W_s^{2/s} W_{s+1}^{2/(s+1)} \cdots W_n^{2/n}) \\
&= E(W_r^{1/r})E(W_{r+1}^{1/(r+1)}) \cdots E(W_s^{2/s})E(W_{s+1})^{2/(s+1)} \cdots E(W_n^{2/n}) \\
&= \prod_{k=r}^{s-1} \frac{1}{(1+1/k)} \prod_{k=s}^{n} \frac{1}{(1+2/k)} \\
&= \frac{r(s+1)}{(n+1)(n+2)}.
\end{aligned} \qquad (8.9)$$

From (8.2), (8.8) and (8.9) we derive the following expressions for covariations and correlation coefficients between the uniform order statistics:

$$\text{Cov}(U_{r,n}, U_{s,n}) = EU_{r,n}U_{s,n} - EU_{r,n}EU_{s,n} = \frac{r(n-s+1)}{(n+1)^2(n+2)}, \quad r \leqslant s, \tag{8.10}$$

and

$$\rho(U_{r,n}, U_{s,n}) = \left(\frac{r(n-s+1)}{s(n-r+1)}\right)^{1/2}. \tag{8.11}$$

It is interesting to note (see, for example, Sathe (1988) and Szekely, Mori (1985)) that for any distribution with a finite second moment, except the uniform distributions,

$$\rho(X_{r,n}, X_{s,n}) < \left(\frac{r(n-s+1)}{s(n-r+1)}\right)^{1/2},$$

that is the equality

$$\rho(X_{r,n}, X_{s,n}) = \left(\frac{r(n-s+1)}{s(n-r+1)}\right)^{1/2}$$

characterizes the family of the uniform distributions.

Exercise 8.3. Find product moments

$$E(U_{r,n}^{\alpha} U_{s,n}^{\beta}), \quad \alpha \geqslant 0, \quad \beta \geqslant 0. \tag{8.12}$$

Exercise 8.4. Let $X_{1,n}, \ldots, X_{n,n}$ be order statistics corresponding to the distribution with the density

$$f(x) = ax^{a-1}, \quad 0 < x < 1, \ a > 0.$$

Find $EX_{r,n}$ and $EX_{r,n}X_{s,n}$, $1 \leqslant r < s \leqslant n$.

Example 8.3. Representation (4.24) is also useful for finding different moments of the uniform order statistics. Denote

$$\mu_k = \frac{v_k}{v_1 + \cdots + v_{n+1}}, \quad k = 1, 2, \ldots, n+1, \tag{8.13}$$

where $v_1, v_2, \ldots$ are independent exponentially $E(1)$ distributed random variables. It follows from (4.24) that

$$(U_{1,n}, U_{2,n}, \ldots, U_{n,n}) \overset{d}{=} (\mu_1, \mu_1 + \mu_2, \ldots, \mu_1 + \mu_2 + \cdots + \mu_n) \tag{8.14}$$

The symmetrical structure of $\mu_1, \ldots, \mu_{n+1}$ and the natural identity

$$\mu_1 + \mu_2 + \cdots + \mu_{n+1} = 1$$

implies that for any $k = 1, 2, \ldots, n+1$,

$$1 = E(\mu_1 + \mu_2 + \cdots + \mu_{n+1}) = (n+1)E\mu_k$$

and

$$E\mu_k = \frac{1}{n+1}.$$

Thus,

$$EU_{k,n} = E(\mu_1 + \mu_2 + \cdots + \mu_k) = \frac{k}{n+1}, \quad k = 1, 2, \ldots, n.$$

Similarly we have equality

$$0 = \text{Var}(\mu_1 + \mu_2 + \cdots + \mu_{n+1}) = \sum_{k=1}^{n+1} \text{Var}(\mu_k) + 2 \sum_{1 \leqslant i < j \leqslant n+1} \text{cov}(\mu_i, \mu_j). \tag{8.15}$$

It is clear that all variances $\text{Var}(\mu_k)$ takes on the same value, say σ^2, as well as all covariances $\text{cov}(\mu_i, \mu_j)$, $i \neq j$, are identical. Let

$$d = \text{cov}(\mu_i, \mu_j), \quad i \neq j.$$

Then we derive from (8.15) that

$$(n+1)\sigma^2 + n(n+1)d = 0.$$

Due to the latter equality we immediately find the expression for correlation coefficients between μ_i and μ_j, $i \neq j$. In fact,

$$\rho(\mu_i, \mu_j) = \frac{\text{cov}(\mu_i, \mu_j)}{(\text{Var}(\mu_i)\text{Var}(\mu_j))^{1/2}} = \frac{d}{\sigma^2} = -\frac{1}{n}, \quad i \neq j. \tag{8.16}$$

It means, in particular, that

$$\rho(U_{1,n}, U_{n,n}) = -\rho(U_{1,n}, 1 - U_{n,n}) = -\rho(\mu_1, \mu_{n+1}) = \frac{1}{n}$$

and this is in agreement with (8.11). To develop further the just-given results we must know the value of σ^2. Recalling that

$$\sigma^2 = \text{Var}(\mu_k), \quad k = 1, 2, \ldots, n+1,$$

and in particular,

$$\sigma^2 = \text{Var}(\mu_1) = \text{Var}(U_{1,n}),$$

we get from (8.8) that

$$\sigma^2 = \frac{n}{(n+1)^2(n+2)}. \tag{8.17}$$

It means that

$$d = \text{cov}(\mu_i, \mu_j) = -\frac{\sigma^2}{n} = -\frac{1}{(n+1)^2(n+2)}, \quad i \neq j. \tag{8.18}$$

Now we can find covariances between different uniform order statistics. Let $i \leqslant j$. Then

$$\operatorname{cov}(U_{i,n}, U_{j,n}) = \operatorname{cov}(\mu_1 + \cdots + \mu_i, \mu_1 + \cdots + \mu_j) = i\sigma^2 + i(j-1)d$$

$$= i\sigma^2 \left(1 - \frac{j-1}{n}\right) = i\frac{n-j+1}{(n+1)^2(n+2)}. \tag{8.19}$$

In particular,

$$\operatorname{Var}(U_{i,n}) = \operatorname{cov}(U_{i,n}, U_{i,n}) = i\frac{n-i+1}{(n+1)^2(n+2)}, \quad i = 1, 2, \ldots, n. \tag{8.20}$$

Indeed, (8.20) and (8.19) coincide with equalities (8.8) and (8.10) respectively.

Thus, we suggested some alternative ways to get moments of the uniform order statistics. Below you will find some exercises, solutions of which are based on representation (4.24).

Exercise 8.5. Find $E(U_{1,n}/(1 - U_{n,n}))^\alpha$.

Exercise 8.6. Find $E(Y/V)$, where

$$Y = U_{r,n}, \quad V = U_{s,n} - U_{m,n} \text{ and } 1 \leqslant m < r < s \leqslant n.$$

Example 8.4. In example 4.7 we introduced differences

$$\delta_1 = U_{1,n} - U_{0,n}, \delta_2 = U_{2,n} - U_{1,n}, \ldots, \delta_n = U_{n,n} - U_{n-1,n}, \delta_{n+1} = U_{n+1,n} - U_{n,n},$$

where $U_{0,n} = 0$, $U_{n+1,n} = 1$, and formed the variational series

$$\delta_{1,n+1} \leqslant \delta_{2,n+1} \leqslant \cdots \leqslant \delta_{n+1,n+1}.$$

We found that the following representation (4.32) is valid for elements of this new variational series:

$$\delta_{k,n+1} \overset{d}{=} \left(\frac{v_1}{n+1} + \frac{v_2}{n} + \cdots + \frac{v_k}{n-k+2}\right)/(v_1 + \cdots + v_{n+1}), \quad k = 1, 2, \ldots, n+1.$$

The technique, used above, enables us to derive that

$$E\delta_{k,n+1} = \frac{1}{n+1}\left(\frac{1}{n+1} + \frac{1}{n} + \cdots + \frac{1}{n-k+2}\right). \tag{8.21}$$

In particular,

$$E\delta_{1,n+1} = \frac{1}{(n+1)^2} \sim \frac{1}{n^2}, \quad n \to \infty,$$

and

$$E\delta_{n+1,n+1} = \frac{1}{n+1}\left(1 + \frac{1}{2} + \cdots + \frac{1}{n+1}\right) \sim \frac{\log n}{n}, \quad n \to \infty.$$

Exercise 8.7. Find variances of order statistics $\delta_{k,n+1}, k = 1, 2, \ldots, n+1$.

8.2 Exponential order statistics

Let $Z_{1,n} \leqslant Z_{2,n} \leqslant \cdots \leqslant Z_{n,n}$, $n = 1, 2, \ldots$, be order statistics corresponding to the standard exponential distribution with d.f.

$$H(x) = 1 - \exp(-x), \quad x > 0.$$

To obtain moments

$$E(Z_{k,n})^{\alpha}, \quad k = 1, 2, \ldots, n,$$

one needs to calculate integrals

$$\frac{n!}{(k-1)!(n-k)!} \int_0^{\infty} x^{\alpha} (H(x))^{k-1} (1 - H(x))^{n-k} dH(x)$$

$$= \frac{n!}{(k-1)!(n-k)!} \int_0^{\infty} x^{\alpha} (1 - e^{-x})^{k-1} e^{-x(n-k+1)} dx$$

$$= \frac{n!}{(k-1)!(n-k)!} \sum_{r=0}^{k-1} (-1)^r \binom{k-1}{r} \int_0^{\infty} x^{\alpha} e^{-x(n-k+r+1)} dx.$$

Since

$$\int_0^{\infty} x^{\alpha} e^{-x(n-k+r+1)} dx = (n-k+r+1)^{-(\alpha+1)} \int_0^{\infty} u^{\alpha} e^{-u} du = \frac{\Gamma(\alpha+1)}{(n-k+r+1)^{(\alpha+1)}},$$

we obtain that

$$E(Z_{k,n})^{\alpha} = \frac{n!}{(k-1)!(n-k)!} \sum_{r=0}^{k-1} (-1)^r \binom{k-1}{r} \frac{\Gamma(\alpha+1)}{(n-k+r+1)^{(\alpha+1)}}. \tag{8.22}$$

For instance, if $k = 1$, then

$$E(Z_{1,n})^{\alpha} = n \frac{\Gamma(\alpha+1)}{n^{(\alpha+1)}} = \frac{\Gamma(\alpha+1)}{n^{\alpha}}, \quad \alpha > -1. \tag{8.23}$$

For $k = 2$ and $\alpha > -1$ we have

$$E(Z_{2,n})^{\alpha} = n(n-1)\Gamma(\alpha+1)\{(n-1)^{-(\alpha+1)} - n^{-(\alpha+1)}\}. \tag{8.24}$$

Some simplifications of the general expression (8.22) are due to representation (4.15).

Example 8.5. Let us recall that the exponential order statistics are expressed in terms of sums of independent random variables:

$$(Z_{1,n}, Z_{2,n}, \ldots, Z_{n,n}) \stackrel{d}{=} \left(\frac{v_1}{n}, \frac{v_1}{n} + \frac{v_2}{n-1}, \ldots, \frac{v_1}{n} + \frac{v_2}{n-1} + \cdots + \frac{v_{n-1}}{2} + v_n \right),$$

where $v_1, v_2, \ldots$ are independent exponential $E(1)$ random variables. Immediately we obtain that

$$EZ_{k,n} = E\left(\frac{v_1}{n} + \frac{v_2}{n-1} + \cdots + \frac{v_k}{n-k+1} \right) = \sum_{r=1}^{k} \frac{1}{n-r+1} \tag{8.25}$$

and

$$\operatorname{Var}(Z_{k,n}) = \sum_{r=1}^{k} \operatorname{Var}\left(\frac{v_r}{n-r+1}\right) = \sum_{r=1}^{k} \frac{1}{(n-r+1)^2}, \tag{8.26}$$

so far as $Ev = \operatorname{Var} v = 1$ if v has the standard exponential distribution. It follows from (8.25) and (8.26) that

$$E(Z_{k,n})^2 = \sum_{r=1}^{k} \frac{1}{(n-r+1)^2} + \left(\sum_{r=1}^{k} \frac{1}{n-r+1}\right)^2. \tag{8.27}$$

Comparing (8.25) and (8.27) with (8.22) (under $\alpha = 1$ and $\alpha = 2$) we derive the following identities:

$$\frac{n!}{(k-1)!(n-k)!} \sum_{r=0}^{k-1} (-1)^r \frac{\binom{k-1}{r}}{(n-k+r+1)^2} = \sum_{r=1}^{k} \frac{1}{n-r+1} \tag{8.28}$$

and

$$\frac{2(n!)}{(k-1)!(n-k)!} \sum_{r=0}^{k-1} (-1)^r \frac{\binom{k-1}{r}}{(n-k+r+1)^3}$$

$$= \sum_{r=1}^{k} \frac{1}{(n-r+1)^2} + \left(\sum_{r=1}^{k} \frac{1}{n-r+1}\right)^2. \tag{8.29}$$

Remark 8.1. It is interesting to see that

$$EZ_{1,n} = \frac{1}{n}$$

and

$$\operatorname{Var} Z_{1,n} = \frac{1}{n^2}$$

tend to zero, as $n \to \infty$, while

$$EZ_{n,n} = \sum_{r=1}^{n} \frac{1}{n-r+1} = \sum_{r=1}^{n} \frac{1}{r} \sim \log n \to \infty, \quad n \to \infty, \tag{8.30}$$

and

$$\operatorname{Var} Z_{n,n} = \sum_{r=1}^{n} \frac{1}{r^2} \to \frac{\pi^2}{6}, \quad n \to \infty. \tag{8.31}$$

Example 8.6. It is possible to find some useful relation for moments of the exponential order statistics. For example, the following identity is valid.

$$E(Z_{i,n}^k) = E(Z_{i-1,n}^k) + \frac{k}{n-i+1} E(Z_{i,n}^{k-1}), \quad k \geqslant 1, \ 2 \leqslant i \leqslant n.$$

In fact,

$$E(Z_{i,n}^{k-1}) = \frac{n!}{(i-1)!\,(n-i)!} \int_0^\infty x^{k-1}(1-e^{-x})^{i-1}e^{-(n-i+1)x}dx$$

Integrating by parts, we obtain

$$E(Z_{i,n}^{k-1}) = \frac{n!}{(i-1)!\,(n-i)!k} \int_0^\infty (n-i+1)x^k(1-e^{-x})^{i-1}e^{-(n-i+1)x}dx$$

$$-(i-1) \int_0^\infty x^k(1-e^{-x})^{i-2}e^{-(n-i+2)x}dx$$

$$= \frac{n-i+1}{k} E(Z_{i,n}^k) - \frac{n-i+1}{k} E(Z_{i-1,n}^k).$$

Thus

$$E(Z_{i,n}^k) = E(Z_{i-1,n}^k) + \frac{k}{n-i+1} E(Z_{i,n}^{k-1}).$$

Exercise 8.8. Find central moments $E(Z_{k,n} - EZ_{k,n})^3$.

Exercise 8.9. Find covariances between order statistics $Z_{r,n}$ and $Z_{s,n}$.

Exercise 8.10. Let $W = aZ_{r,n} + bZ_{s,n}$, where $r < s$. Find the variance of W.

Check your solutions

Exercise 8.1 (answer).

$$\mathrm{Var}\left(\frac{1}{U_{k,n}}\right) = \frac{n(n-k+1)}{(k-1)^2(k-2)}, \quad 3 \leqslant k \leqslant n.$$

Exercise 8.2 (answer).

$$E(U_{k,n} - EU_{k,n})^3 = \frac{2k(n-k+1)(n-2k+1)}{(n+1)^3(n+2)(n+3)}, \quad 1 \leqslant k \leqslant n.$$

Exercise 8.3 (answer).

$$E(U_{r,n}^\alpha U_{s,n}^\beta) = \frac{n!\,\Gamma(r+\alpha)\Gamma(s+\alpha+\beta)}{(r-1)!\,\Gamma(s+\alpha)\Gamma(n+1+\alpha+\beta)}.$$

Exercise 8.4 (hint and answer). Use the relation

$$X_{r,n} \overset{d}{=} (U_{r,n})^{1/a},$$

which expresses $X_{r,n}$ via the uniform order statistics and the result of exercise 8.3 to get equalities

$$EX_{r,n} = \frac{n!\,\Gamma(r+1/a)}{(r-1)!\,\Gamma(n+1+1/a)}$$

and

$$EX_{r,n}X_{s,n} = E(U_{r,n})^{1/a}(U_{s,n})^{1/a} = \frac{n!\,\Gamma(r+1/a)\Gamma(s+2/a)}{(r-1)!\,\Gamma(s+1/a)\Gamma(n+1+2/a)}.$$

Exercise 8.5 (solution). Taking into account representation (4.24), we have relation

$$\frac{U_{1,n}}{1-U_{n,n}} \overset{d}{=} \frac{v_1}{v_{n+1}},$$

where v_1 and v_{n+1} are independent and have the standard $E(1)$ exponential distribution. Then

$$E\left(\frac{U_{1,n}}{1-U_{n,n}}\right)^\alpha = E\left(\frac{v_1}{v_{n+1}}\right)^\alpha = Ev^\alpha Ev^{-\alpha},$$

where v has density function $\exp(-x)$, $x > 0$. Note that

$$Ev^\beta = \int_0^\infty x^\beta e^{-x} dx$$

is finite only if $\beta > -1$ and

$$Ev^\beta = \Gamma(\beta + 1)$$

for $\beta > -1$. Hence

$$E\left(\frac{U_{1,n}}{1-U_{n,n}}\right)^\alpha = \begin{cases} \infty, & \text{if } |\alpha| \geqslant 1, \\ \Gamma(\alpha+1)\Gamma(1-\alpha), & \text{if } |\alpha| < 1. \end{cases}$$

We can recall some properties of gamma functions, such as

$$\Gamma(\alpha+1) = \alpha\Gamma(\alpha)$$

and

$$\Gamma(\alpha)\Gamma(1-\alpha) = \frac{\pi}{\sin \pi\alpha}$$

for $\alpha > 0$, and then derive that

$$E\left(\frac{U_{1,n}}{1-U_{n,n}}\right)^\alpha = \frac{\alpha\pi}{\sin \pi\alpha}$$

for $1 > \alpha > 0$.
Similarly,

$$E\left(\frac{U_{1,n}}{1-U_{n,n}}\right)^\alpha = \Gamma(\alpha+1)\Gamma(1-\alpha) = (-\alpha)\Gamma(\alpha+1)\Gamma(-\alpha) = \frac{\alpha\pi}{\sin \pi\alpha}$$

for $-1 < \alpha < 0$.

Exercise 8.6 (solution). Coming back to representation (4.24) we derive that

$$\frac{Y}{V} \overset{d}{=} \frac{v_1 + \cdots + v_r}{v_{m+1} + \cdots + v_r + v_{r+1} + \cdots + v_s} = \frac{v_1 + \cdots + v_m}{v_{m+1} + \cdots + v_s} + \frac{v_{m+1} + \cdots + v_r}{v_{m+1} + \cdots + v_s}.$$

Hence,

$$E\left(\frac{Y}{V}\right) = E(v_1 + \cdots + v_m)E\left(\frac{1}{v_{m+1} + \cdots + v_s}\right) + E\left(\frac{v_{m+1} + \cdots + v_r}{v_{m+1} + \cdots + v_s}\right).$$

Note that

$$E\left(\frac{v_{m+1}+\cdots+v_r}{v_{m+1}+\cdots+v_s}\right) = EU_{r-m,s-m-1} = \frac{r-m}{s-m}$$

and

$$E(v_1+\cdots+v_m) = m.$$

Consider now the sum $v_{m+1}+\cdots+v_s$, which has the gamma distribution with parameter $(s-m)$. Hence (due to the fact that $s-m > 1$) we obtain that

$$E\left(\frac{1}{v_m+\cdots+v_s}\right) = \frac{1}{(s-m-1)!}\int_0^\infty x^{s-m-2}e^{-x}dx = \frac{\Gamma(s-m-1)}{\Gamma(s-m)} = \frac{1}{s-m-1}.$$

Finally we get

$$E\left(\frac{Y}{V}\right) = \frac{m}{s-m-1} + \frac{r-m}{s-m}.$$

Exercise 8.7 (solution). Comparing relations (4.24), (4.32) and (8.14) we see that

$$\delta_{k,n+1} \stackrel{d}{=} \frac{\mu_1}{n+1} + \frac{\mu_2}{n} + \cdots + \frac{\mu_k}{n-k+2}, \quad k = 1,2,\ldots,n+1,$$

where $\mu_1, \mu_2,\ldots,\mu_{n+1}$ are defined in (8.13). The variance of $\delta_{k,n+1}$ is expressed via variances of μ's, which are equal to

$$\sigma^2 = \frac{n}{(n+1)^2(n+2)}$$

(see (8.17)), and covariances $\mathrm{cov}\,(\mu_i,\mu_j)$, $i \neq j$, which (as we know from (8.18)) are identical:

$$d = -\frac{\sigma^2}{n} = -\frac{1}{(n+1)^2(n+2)}.$$

Really, we have

$$\mathrm{Var}\,(\delta_{k,n+1}) = \mathrm{Var}\left(\frac{\mu_1}{n+1} + \frac{\mu_2}{n} + \cdots + \frac{\mu_k}{n-k+2}\right)$$

$$= \sum_{i=1}^k \mathrm{Var}\left(\frac{\mu_i}{n-i+2}\right) + 2\sum_{1\leqslant i<j\leqslant k} \mathrm{cov}\left(\frac{\mu_i}{n-i+2}, \frac{\mu_j}{n-j+2}\right)$$

$$= \sigma^2 \sum_{i=1}^k \frac{1}{(n-i+2)^2} + 2d \sum_{1\leqslant i<j\leqslant k} \frac{1}{(n-i+2)(n-j+2)}$$

$$= \frac{n}{(n+1)^2(n+2)} \sum_{i=1}^k \frac{1}{(n-i+2)^2} - \frac{2}{(n+1)^2(n+2)} \sum_{1\leqslant i<j\leqslant k} \frac{1}{(n-i+2)(n-j+2)},$$

for $k = 1,2,\ldots,n+1$.

In particular,

$$\mathrm{Var}\,(\delta_{1,n+1}) = \frac{n}{(n+1)^4(n+2)}$$

and

$$\mathrm{Var}\,(\delta_{2,n+1}) = \frac{n}{(n+1)^4(n+2)} + \frac{1}{n(n+1)^2(n+2)} - \frac{2}{n(n+1)^3(n+2)}.$$

Exercise 8.8 (solution). Recalling (4.15) one gets that

$$E(Z_{k,n} - EZ_{k,n})^3 = E\left(\frac{v_1 - 1}{n} + \frac{v_2 - 1}{n-1} + \cdots + \frac{v_k - 1}{n-k+1}\right)^3$$

$$= \sum_{r=1}^{k} E\left(\frac{v_r - 1}{n-r+1}\right)^3$$

$$= \sum_{r=1}^{k} \frac{1}{(n-r+1)^3} E(v-1)^3,$$

where v has the standard $E(1)$ exponential distribution. We have also that

$$E(v-1)^3 = Ev^3 - 3Ev^2 + 3Ev - 1 = \Gamma(4) - 3\Gamma(3) + 3 - 1 = 2.$$

Hence,

$$E(Z_{k,n} - EZ_{k,n})^3 = 2\sum_{r=1}^{k} \frac{1}{(n-r+1)^3}.$$

Exercise 8.9 (solution). Let $r \leqslant s$.

Since sums

$$\frac{v_1}{n} + \frac{v_2}{n-1} + \cdots + \frac{v_r}{n-r+1}$$

and

$$\frac{v_{r+1}}{n-r} + \cdots + \frac{v_s}{n-s+1}$$

are independent, we get that

$$\text{Cov}(Z_{r,n}, Z_{s,n}) = \text{Cov}\left(\frac{v_1}{n} + \frac{v_2}{n-1} + \cdots + \frac{v_r}{n-r+1}, \frac{v_1}{n} + \frac{v_2}{n-1} + \cdots + \frac{v_s}{n-s+1}\right)$$

$$= \text{Cov}\left(\frac{v_1}{n} + \frac{v_2}{n-1} + \cdots + \frac{v_r}{n-r+1}, \frac{v_1}{n} + \frac{v_2}{n-1} + \cdots + \frac{v_r}{n-r+1}\right)$$

$$= \text{Var}\left(\frac{v_1}{n} + \frac{v_2}{n-1} + \cdots + \frac{v_r}{n-r+1}\right) = \text{Var}\, Z_{r,n} = \sum_{k=1}^{r} \frac{1}{(n-k+1)^2}.$$

Exercise 8.10 (solution). By virtue of (4.15) we get that

$$W \stackrel{d}{=} a\left(\frac{v_1}{n} + \frac{v_2}{n-1} + \cdots + \frac{v_r}{n-r+1}\right) + b\left(\frac{v_1}{n} + \frac{v_2}{n-1} + \cdots + \frac{v_s}{n-s+1}\right)$$

$$= (a+b)\left(\frac{v_1}{n} + \frac{v_2}{n-1} + \cdots + \frac{v_r}{n-r+1}\right) + b\left(\frac{v_{r+1}}{n-r} + \frac{v_2}{n-1} + \cdots + \frac{v_s}{n-s+1}\right).$$

The independence of summands enables us to find the variance of the sum:

$$\text{Var}\, W = (a+b)^2 \sum_{k=1}^{r} \frac{1}{(n-k+1)^2} + b^2 \sum_{k=r+1}^{s} \frac{1}{(n-k+1)^2}.$$

Chapter 9

Moment Relations for Order Statistics: Normal Distribution

Below we investigate some useful relations for different moment characteristics of order statistics based on the samples from normal distributions.

Let $X_1, X_2, \ldots$ be independent random variables having the standard normal distribution and $X_{1,n} \leqslant \cdots \leqslant X_{n,n}$ be the corresponding normal order statistics. It is known that the normal distribution is the most popular in the mathematical statistics. Statistical methods related to normal samples are deeply investigated and have a very long history. However, there are some problems when we want to calculate moments of normal order statistics. Indeed, one can write immediately (remembering the general form for moments of order statistics) that

$$EX_{k,n}^r = \frac{n!}{(k-1)!(n-k)!} \int_{-\infty}^{\infty} x^r \Phi^{k-1}(x)(1-\Phi(x))^{n-k}\varphi(x)dx, \qquad (9.1)$$

where

$$\varphi(x) = \frac{1}{\sqrt{2\pi}} \exp\left(-\frac{x^2}{2}\right)$$

and

$$\Phi(x) = \int_{-\infty}^{x} \varphi(t)dt.$$

There are effective numerical methods to compute integrals (9.1). Unlike the cases of the uniform and exponential order statistics, moments (9.1) have the explicit expressions only in some special situations for small sample sizes.

Example 9.1. Consider the case $n = 2$. We get that

$$EX_{2,2} = 2\int_{-\infty}^{\infty} x\Phi(x)\varphi(x)dx = -2\int_{-\infty}^{\infty} \Phi(x)d(\varphi(x))$$

$$= 2\int_{-\infty}^{\infty} \varphi^2(x)dx = \frac{1}{\pi}\int_{-\infty}^{\infty} \exp(-x^2)dx = \frac{1}{\sqrt{\pi}}.$$

M. Ahsanullah et al., *An Introduction to Order Statistics*,
Atlantis Studies in Probability and Statistics 3,
DOI: 10.2991/978-94-91216-83-1_9, © Atlantis Press 2013

From the identity

$$E(X_{1,2} + X_{2,2}) = E(X_1 + X_2) = 0$$

we obtain now that

$$EX_{1,2} = -EX_{2,2} = -\frac{1}{\sqrt{\pi}}.$$

Remark 9.1. If we have two samples $X_1, X_2, \ldots, X_n$ (from the standard $N(0,1)$ normal distribution) and $Y_1, Y_2, \ldots, Y_n$ (from the normal $N(a, \sigma^2)$ distribution with expectation a and variance $\sigma^2, \sigma > 0$), then evidently

$$E(Y_{k,n}) = a + \sigma E(X_{k,n}).$$

Remark 9.2. Any normal $N(a, \sigma^2)$ distribution is symmetric with respect to its expectation a. Hence, it is easy to see that for order statistics $Y_{1,n} \leqslant \cdots \leqslant Y_{n,n}$ the following relations are true:

$$EY_{k,n} = 2a - EY_{n-k+1,n}, \quad k = 1, 2, \ldots, n; \tag{9.2}$$

$$E(Y_{k,n} - a)^m = (-1)^m E(Y_{n-k+1,n} - a)^m, \quad k = 1, 2, \ldots, n; \quad m = 1, 2, \ldots. \tag{9.3}$$

It follows from (9.3) that if $Y_{k+1,2k+1}$ is a sample median from the normal $N(a, \sigma^2)$ distribution, then

$$E(Y_{k+1,2k+1} - a)^{2r-1} = 0, \quad r = 1, 2, \ldots, \tag{9.4}$$

and, in particular,

$$EY_{k+1,2k+1} = a. \tag{9.5}$$

Exercise 9.1. Let $X_{1,3} \leqslant X_{2,3} \leqslant X_{3,3}$ be the order statistics corresponding to the standard normal distribution. Find $EX_{1,3}$, $EX_{2,3}$ and $EX_{3,3}$.

Exercise 9.2. Let $X_{1,3} \leqslant X_{2,3} \leqslant X_{3,3}$ be the order statistics corresponding to the standard normal distribution. Find $E(X_{k,3})^2$ and $\text{Var}(X_{k,3})$, $k = 1, 2, 3$.

The just-obtained results show that the explicit expressions for moments of the normal order statistics are rather complicated, although the normal distribution possesses a number of useful properties, which can simplify the computational schemes.

Example 9.2. A lot of statistical procedures for the normal distribution are based on the independence property of vector

$$(X_1 - \overline{X}, X_2 - \overline{X}, \ldots, X_n - \overline{X})$$

and the sample mean

$$\overline{X} = \frac{X_1 + X_2 + \cdots + X_n}{n}.$$

What is more important for us, this yields the independence of vector

$$(X_{1,n} - \overline{X}, X_{2,n} - \overline{X}, \ldots, X_{n,n} - \overline{X})$$

and the sample mean $\overline{X}$.

Let $X_1, X_2, \ldots, X_n$ be a sample from the standard normal distribution. We see that then

$$E(X_{k,n} - \overline{X})\overline{X} = E(X_{k,n} - \overline{X})E\overline{X} = 0 \tag{9.6}$$

and we obtain the following results:

$$EX_{k,n}\overline{X} = E(\overline{X}^2) = \operatorname{Var}\overline{X} = \frac{1}{n}, \quad k = 1, 2, \ldots, n, \tag{9.7}$$

and hence

$$\sum_{m=1}^{n} EX_{k,n}X_m = 1, \tag{9.8}$$

as well as

$$\sum_{m=1}^{n} EX_{k,n}X_{m,n} = 1. \tag{9.9}$$

As corollaries of (9.8) we get that

$$E(X_{k,n}X_m) = \frac{1}{n} \tag{9.10}$$

and

$$\operatorname{cov}(X_{k,n}, X_m) = E(X_{k,n}X_m) - EX_{k,n}EX_m = \frac{1}{n} \tag{9.11}$$

for any $k = 1, 2, \ldots, n$, $m = 1, 2, \ldots, n$ and $n = 1, 2, \ldots$.

Similarly to (9.6), we can get also that

$$\operatorname{cov}(X_{k,n} - \overline{X}, \overline{X}) = 0$$

and hence

$$\operatorname{cov}(X_{k,n}, \overline{X}) = \operatorname{cov}(\overline{X}, \overline{X}) = \operatorname{Var}(\overline{X}) = \frac{1}{n}. \tag{9.12}$$

Since at the same time

$$n\overline{X} = (X_1 + X_2 + \cdots + X_n)$$

and

$$n\overline{X} = (X_{1,n} + X_{2,n} + \cdots + X_{n,n}), \tag{9.13}$$

one obtains from (9.12) that

$$\sum_{m=1}^{n} \operatorname{cov}(X_{k,n}, X_{m,n}) = 1, \quad k = 1, 2, \ldots, n, \tag{9.14}$$

and

$$\operatorname{cov}(X_{k,n}, X_m) = \frac{1}{n}, \quad 1 \leqslant k \leqslant n, \ 1 \leqslant m \leqslant n.$$

The symmetry of the normal distribution with respect to its expectation also gives some simplifications.

Example 9.3. Let again $X_{1,n}, X_{2,n}, \ldots, X_{n,n}$ be order statistics from the standard normal distribution. The symmetry of the normal distribution implies that

$$(X_{1,n}, X_{2,n}, \ldots, X_{n,n}) \overset{d}{=} (-X_{n,n}, -X_{n-1,n}, \ldots, -X_{1,n}). \tag{9.15}$$

Hence

$$EX_{k,n} = -EX_{n-k+1,n}, \quad k = 1, 2, \ldots, n, \tag{9.16}$$

$$E(X_{k,n})^2 = E(X_{n-k+1,n})^2, \quad k = 1, 2, \ldots, n, \tag{9.17}$$

$$\operatorname{Var}(X_{k,n}) = \operatorname{Var}(X_{n-k+1,n}), \quad k = 1, 2, \ldots, n. \tag{9.18}$$

It follows also from (9.15) that

$$E(X_{r,n} X_{s,n}) = E(X_{n-r+1,n} X_{n-s+1,n}) \tag{9.19}$$

and

$$\operatorname{cov}(X_{r,n}, X_{s,n}) = \operatorname{cov}(X_{n-r+1,n}, X_{n-s+1,n}) \tag{9.20}$$

for any $1 \leqslant r, s \leqslant n$.

Exercise 9.3. Let $X_{1,3} \leqslant X_{2,3} \leqslant X_{3,3}$ be the order statistics corresponding to the standard normal distribution. Find covariances

$$\operatorname{cov}(X_{r,3}, X_{s,3})$$

for $1 \leqslant r < s \leqslant 3$.

In some special procedures we need to find moments of maximal or minimal values of dependent normal random variables.

Example 9.4. Let the initial random variables X_1, X_2 have jointly the bivariate normal

$$N(a_1, a_2, \sigma_1^2, \sigma_2^2, \rho)$$

distribution. This means that

$$EX_1 = a_1, \quad EX_2 = a_2, \quad \operatorname{Var} X_1 = \sigma_1^2, \quad \operatorname{Var} X_2 = \sigma_2^2$$

and the correlation coefficient between X_1 and X_2 equals ρ. Let us find $EX_{1,2}$ and $EX_{2,2}$. We have evidently the following equalities:

$$X_{2,2} = \max\{X_1, X_2\} = X_1 + \max\{0, X_2 - X_1\}$$

and

$$EX_{2,2} = EX_1 + E\max\{0, Z\} = a_1 + E\max\{0, Z\},$$

where Z denotes the difference $X_2 - X_1$, which also has the normal distribution with expectation

$$b = EZ = E(X_2 - X_1) = a_2 - a_1$$

and variance

$$\sigma^2 = \operatorname{Var} Z = \operatorname{Var} X_1 + \operatorname{Var} X_2 - 2\operatorname{cov}\{X_1, X_2\} = \sigma_1^2 + \sigma_2^2 - 2\rho\sigma_1\sigma_2.$$

If $\sigma^2 = 0$, i.e. $Z = X - Y$ has the degenerate distribution, then

$$EX_{2,2} = a_1 + \max\{0, b\} = \max\{a_1, a_2\}$$

and

$$EX_{1,2} = E(X_1 + X_2) - EX_{2,2} = a_1 + a_2 - \max\{a_1, a_2\} = \min\{a_1, a_2\}.$$

Consider now the case $\sigma^2 > 0$. In this situation

$$EX_{2,2} = a_1 + E\max\{0, b + \sigma Y\},$$

where Y has the standard $N(0,1)$ normal distribution with the density function

$$\varphi(x) = \frac{1}{\sqrt{2\pi}}\exp\left(-\frac{x^2}{2}\right)$$

and distribution function

$$\Phi(x) = \int_{-\infty}^{x}\varphi(t)dt.$$

It is not difficult to see that

$$E\max\{0, b + \sigma Y\} = b + E\max\{-b, \sigma Y\} = b + \sigma E\max\{c, Y\},$$

where

$$c = -\frac{b}{\sigma}.$$

Further,

$$E \max\{c,Y\} = cP\{Y \leqslant c\} + \int_c^\infty x\varphi(x)dx = c\Phi(c) - \int_c^\infty d\varphi(x) = c\Phi(c) + \varphi(c).$$

We have thus, that

$$EX_{2,2} = a_1 + E \max\{0, b + \sigma Y\} = a_1 + b + \sigma E \max\{c, Y\}$$
$$= a_1 + b + \sigma(c\Phi(c) + \varphi(c)) = a_2 + \sigma(c\Phi(c) + \varphi(c))$$

and

$$EX_{1,2} = E(X_1 + X_2) - EX_{2,2} = a_1 - \sigma(c\Phi(c) + \varphi(c)).$$

In the partial case, when $a_1 = a_2 = a$, we have the equalities

$$EX_{2,2} = a + \sigma\varphi(0) = a + \frac{\sigma}{\sqrt{2\pi}}$$

and

$$EX_{1,2} = a - \frac{\sigma}{\sqrt{2\pi}}.$$

We considered some situations when it is possible to get the exact expressions for moments of order statistics from the normal populations. Note that there are different tables (see, for example, Teichroew (1956) or Tietjen, Kahaner and Beckman (1977)) which give expected values and some other moments of order statistics for samples of large sizes from the normal distribution.

Check your solutions

Exercise 9.1 (solution). It follows from remark 9.2 that $EX_{2,3} = 0$ and $EX_{1,3} = -EX_{3,3}$. Thus, we need to find $EX_{3,3}$ only. We see that

$$EX_{3,3} = 3\int_{-\infty}^\infty x\Phi^2(x)\varphi(x)dx = 3\int_{-\infty}^\infty \Phi^2(x)d\varphi(x) = 6\int_{-\infty}^\infty \Phi(x)\varphi^2(x)dx.$$

Consider

$$I(a) = \int_{-\infty}^\infty \Phi(ax)\varphi^2(x)dx.$$

We obtain that

$$I(0) = \frac{1}{2}\int_{-\infty}^\infty \varphi^2(x)dx = \frac{1}{4\pi}\int_{-\infty}^\infty \exp(-x^2)dx = \frac{1}{(4\sqrt{\pi})}$$

and

$$I'(a) = \int_{-\infty}^\infty x\varphi(ax)\varphi^2(x)dx =$$

$$\frac{1}{(2\pi)^{3/2}} \int_{-\infty}^{\infty} x \exp\{-X^2(a^2+2)\}dx = 0$$

It means that

$$I(a) = \frac{1}{(4\sqrt{\pi})}$$

and, in particular,

$$\int_{-\infty}^{\infty} \Phi(ax)\varphi^2(x)dx = I(1) = \frac{1}{(4\sqrt{\pi})}.$$

Finally, we have that

$$EX_{3,3} = 6I(1) = \frac{3}{2\sqrt{\pi}}.$$

Exercise 9.2 (solution). Due to symmetry of the standard normal distribution,

$$E(X_{1,3})^2 = E(X_{3,3})^2$$

and

$$\operatorname{Var} X_{1,3} = \operatorname{Var} X_{3,3}.$$

We have also

$$
\begin{aligned}
E(X_{3,3})^2 &= 3\int_{-\infty}^{\infty} x^2 \Phi^2(x)\varphi(x)dx \\
&= -3\int_{-\infty}^{\infty} x\Phi^2(x)d\varphi(x) \\
&= 3\int_{-\infty}^{\infty} \varphi(x)d(x\Phi^2(x)) \\
&= 3\int_{-\infty}^{\infty} \varphi(x)\Phi^2(x)dx + 6\int_{-\infty}^{\infty} x\varphi^2(x)\Phi(x)dx \\
&= \int_{-\infty}^{\infty} d(\Phi^3(x)) + \frac{3}{\pi}\int_{-\infty}^{\infty} x\exp(-x^2)\Phi(x)dx \\
&= 1 - \frac{3}{2\pi}\int_{-\infty}^{\infty} \Phi(x)d(\exp(-x^2)) \\
&= 1 + \frac{3}{2\pi}\int_{-\infty}^{\infty} \exp(-x^2)\varphi(x)dx \\
&= 1 + \frac{3}{(2\pi)^{3/2}}\int_{-\infty}^{\infty} \exp\left(-\frac{3x^2}{2}\right)dx \\
&= 1 + \frac{\sqrt{3}}{2\pi}.
\end{aligned}
$$

Taking into account that

$$EX_{3,3} = \frac{3}{2\sqrt{\pi}}$$

(see exercise 9.1), we obtain that

$$\text{Var}(X_{3,3}) = E(X_{3,3})^2 - (EX_{3,3})^2 = 1 + \frac{\sqrt{3}}{2\pi} - \frac{9}{4\pi}.$$

Further,

$$E(X_{2,3})^2 = 6\int_{-\infty}^{\infty} x^2\Phi(x)(1-\Phi(x))\varphi(x)dx$$

$$= 6\int_{-\infty}^{\infty} x^2\Phi(x)\varphi(x)dx - 6\int_{-\infty}^{\infty} x^2\Phi^2(x)\varphi(x)dx$$

$$= 6\int_{-\infty}^{\infty} x^2\Phi(x)\varphi(x)dx - 2E(X_{3,3})^2.$$

We obtain that

$$\int_{-\infty}^{\infty} x^2\Phi(x)\varphi(x)dx = -\int_{-\infty}^{\infty} x\Phi(x)d\varphi(x)$$

$$= \int_{-\infty}^{\infty} \varphi(x)d(x\Phi(x))$$

$$= \int_{-\infty}^{\infty} \varphi(x)\Phi(x)dx + \int_{-\infty}^{\infty} x\varphi^2(x)dx$$

$$= \frac{1}{2}\int_{-\infty}^{\infty} d(\Phi^2(x)) + \frac{1}{2\pi}\int_{-\infty}^{\infty} x\exp(-x^2)dx = \frac{1}{2}.$$

Hence,

$$E(X_{2,3})^2 = 3 - 2E(X_{3,3})^2 = 3 - 2\left(1 + \frac{\sqrt{3}}{2\pi}\right) = 1 - \frac{\sqrt{3}}{\pi}$$

and

$$\text{Var}(X_{2,3}) = E(X_{2,3})^2 - (EX_{2,3})^2 = 1 - \frac{\sqrt{3}}{\pi},$$

so far as $EX_{2,3} = 0$.

Thus,

$$E(X_{1,3})^2 = E(X_{3,3})^2 = 1 + \frac{\sqrt{3}}{2\pi},$$

$$E(X_{2,3})^2 = 1 - \frac{\sqrt{3}}{\pi},$$

$$\text{Var}(X_{1,3}) = \text{Var}(X_{3,3}) = 1 + \frac{\sqrt{3}}{2\pi} - \frac{9}{4\pi}$$

and

$$\text{Var}(X_{2,3}) = E(X_{2,3})^2 - (EX_{2,3})^2 = 1 - \frac{\sqrt{3}}{\pi}.$$

Let us note that

$$E(X_{1,3})^2 + E(X_{2,3})^2 + E(X_{3,3})^2 = E(X_1)^2 + (EX_2)^2 + (EX_3)^2 = 3.$$

Exercise 9.3 (solution). It follows from (9.20) that

$$\mathrm{cov}\,(X_{1,3}, X_{2,3}) = \mathrm{cov}\,(X_{2,3}, X_{3,3}).$$

From (9.14) we have also that

$$\mathrm{cov}\,(X_{1,3}, X_{2,3}) + \mathrm{cov}\,(X_{2,3}, X_{2,3}) + \mathrm{cov}\,(X_{2,3}, X_{3,3}) = 1.$$

On combining these relations we get equalities

$$\mathrm{cov}\,(X_{1,3}, X_{2,3}) = \mathrm{cov}\,(X_{2,3}, X_{3,3}) = \frac{1 - \mathrm{cov}\,(X_{2,3}, X_{2,3})}{2} = \frac{1 - \mathrm{Var}\,(X_{2,3})}{2}.$$

It was found in exercise 9.2 that

$$\mathrm{Var}\,(X_{2,3}) = 1 - \frac{\sqrt{3}}{\pi}.$$

Finally,

$$\mathrm{cov}\,(X_{1,3}, X_{2,3}) = \mathrm{cov}\,(X_{2,3}, X_{3,3}) = \frac{\sqrt{3}}{2\pi}.$$

Now we again use (9.14):

$$\mathrm{cov}\,(X_{1,3}, X_{1,3}) + \mathrm{cov}\,(X_{1,3}, X_{2,3}) + \mathrm{cov}\,(X_{1,3}, X_{3,3}) = 1$$

and have the equality

$$\mathrm{cov}\,(X_{1,3}, X_{3,3}) = 1 - \mathrm{cov}\,(X_{1,3}, X_{2,3}) - \mathrm{Var}\,(X_{1,3}).$$

It was found in exercise 9.2 that

$$\mathrm{Var}\,(X_{1,3}) = 1 + \frac{\sqrt{3}}{2\pi} - \frac{9}{4\pi}.$$

Thus,

$$\mathrm{cov}\,(X_{1,3}, X_{3,3}) = 1 - \frac{\sqrt{3}}{2\pi} - \left(1 + \frac{\sqrt{3}}{2\pi} - \frac{9}{4\pi}\right) = \frac{9}{4\pi} - \frac{\sqrt{3}}{\pi} = \frac{9 - 4\sqrt{3}}{4\pi}.$$

Chapter 10

Asymptotic Behavior of the Middle and Intermediate Order Statistics

We begin to study different limit theorems for order statistics. In this chapter we consider the asymptotic distributions for the so-called middle and intermediate order statistics.

It turns out that one, who wants to study asymptotic distributions of suitably normalized and centered order statistics $X_{k,n}$, must distinguish three different options. Since we consider the case, when n (the size of the sample) tends to infinity, it is natural that $k = k(n)$ is a function of n. The order statistics $X_{k(n),n}$ are said to be extreme if $k = k(n)$ or $n - k(n) + 1$ is fixed, as $n \to \infty$. If

$$0 < \liminf_{n \to \infty} \frac{k(n)}{n} \leqslant \limsup_{n \to \infty} \frac{k(n)}{n} < 1,$$

then order statistics $X_{k(n),n}$ are said to be middle. At last, the case when $k(n) \to \infty$, $k(n)/n \to 0$ or $n - k(n) \to \infty$, $k(n)/n \to 1$, corresponds to the so-called intermediate order statistics.

Exercise 10.1. Show that sample quantiles of order p, $0 < p < 1$, present middle order statistics.

The opportunity to express the uniform and exponential order statistics via sums of independent terms enables us to use limit theorems for sums of independent random variables. Really we need to know the following Lyapunov theorem.

Theorem 10.1. *Let $X_1, X_2, \ldots$ be independent random variables with expectations $a_k = EX_k$, variances σ_k^2 and finite moments $\gamma_k = E|X_k - a_k|^3$, $k = 1, 2, \ldots$. Denote*

$$S_n = \sum_{k=1}^{n} (X_k - a_k),$$

$$B_n^2 = \mathrm{Var}\,(S_n) = \sum_{k=1}^{n} \sigma_k^2$$

M. Ahsanullah et al., *An Introduction to Order Statistics*,
Atlantis Studies in Probability and Statistics 3,
DOI: 10.2991/978-94-91216-83-1_10, © Atlantis Press 2013

and

$$L_n = \sum_{k=1}^{n} \frac{\gamma_k}{B_n^3}.$$ (10.1)

If Lyapunov ratio (10.1) *converges to zero, as* $n \to \infty$, *then*

$$\sup_x |P\{S_n/B_n < x\} - \Phi(x)| \to 0,$$ (10.2)

where

$$\Phi(x) = \frac{1}{\sqrt{2\pi}} \int_{-\infty}^{x} \exp\left(-\frac{t^2}{2}\right) dt$$

is the d.f. of the standard normal distribution.

Example 10.1. Consider a sequence of exponential order statistics $Z_{k(n),n}$, $n = 1, 2, \ldots$, where $k(n) \to \infty$ and $\limsup k(n)/n < 1$, as $n \to \infty$. As we know from chapter 4, the following relation is valid for the exponential order statistics:

$$Z_{k,n} \overset{d}{=} \frac{\nu_1}{n} + \frac{\nu_2}{n-1} + \cdots + \frac{\nu_k}{n-k+1}, \quad k = 1, 2, \ldots, n,$$

where $\nu_1, \nu_2, \ldots$ is a sequence of independent identically distributed random variables, having the standard exponential $E(1)$ distribution. Let us check if the Lyapunov theorem can be used for independent terms

$$X_k = \frac{\nu_k}{n-k+1}.$$

We obtain easily that

$$a_k = EX_k = \frac{1}{n-k+1},$$

$$\sigma_k^2 = \mathrm{Var}(X_k) = \frac{1}{(n-k+1)^2}$$

and

$$\gamma_k = \frac{\gamma}{(n-k+1)^3},$$

where

$$\gamma = E|\nu - 1|^3 = \int_0^1 (1-x)^3 e^{-x} dx + \int_1^\infty (x-1)^3 e^{-x} dx = \frac{12}{e} - 2.$$ (10.3)

In our case

$$B_{k,n}^2 = \mathrm{Var}(Z_{k,n}) = \sum_{r=1}^{k} \sigma_r^2 = \sum_{r=1}^{k} \frac{1}{(n-r+1)^2} = \sum_{r=n-k+1}^{n} \frac{1}{r^2}$$ (10.4)

and

$$\Gamma_{k,n} = \sum_{r=1}^{k} E|X_r - a_r|^3 = \gamma \sum_{r=1}^{k} \frac{1}{(n-r+1)^3} = \gamma \sum_{r=n-k+1}^{n} \frac{1}{r^3}.$$ (10.5)

The restriction $\limsup k(n)/n < 1$, as $n \to \infty$, means that there exists such p, $0 < p < 1$, that $k \leqslant pn$, for all sufficiently large n. For such values n we evidently have the following inequalities:

$$B_{k,n}^2 \geqslant \frac{k}{n^2}, \quad \Gamma_{k,n} \leqslant \frac{k\gamma}{(n(1-p))^3}, \quad L_{k,n} = \frac{\Gamma_{k,n}}{B_{k,n}^3} \leqslant \frac{\gamma}{(1-p)^3 k^{1/2}}. \qquad (10.6)$$

Hence, if $\limsup k(n)/n < 1$ and $k(n) \to \infty$, as $n \to \infty$, then the Lyapunov ratio $L_{k(n),n}$ tends to zero and this provides the asymptotic normality of

$$\frac{Z_{k(n),n} - A_{k,n}}{B_{k,n}},$$

where

$$a_{k,n} = \sum_{r=1}^{k} \frac{1}{n-r+1}.$$

Remark 10.1. It is not difficult to see that example 10.1 covers the case of sample quantiles $Z_{[pn]+1,n}$, $0 < p < 1$. Moreover if $k(n) \sim pn$, $0 < p < 1$, as $n \to \infty$, then

$$a_{k,n} = \sum_{r=1}^{k} \frac{1}{n-r+1} = \sum_{r=n-k+1}^{n} \frac{1}{r} \sim -\log(1-p), \quad n \to \infty,$$

and

$$B_{k,n}^2 = \sum_{r=n-k+1}^{n} \frac{1}{r^2} \sim \int_{n-k+1}^{n} x^{-2}dx \sim \frac{p}{(1-p)n},$$

as $n \to \infty$.

All the saying with some additional arguments enable us to show that for any $0 < p < 1$, as $n \to \infty$,

$$\sup_{x} |P\{(Z_{[pn]+1,n} + \log(1-p))(n(1-p)/p)^{1/2} < x\} - \Phi(x)| \to 0. \qquad (10.7)$$

In particular, the following relation is valid for the exponential sample median:

$$\sup_{x} |P\{n^{1/2}(Z_{[n/2]+1,n} - \log 2) < x\} - \Phi(x)| \to 0, \qquad (10.8)$$

as $n \to \infty$.

Remark 10.2. More complicated and detailed arguments allow to prove that suitably centered and normalized exponential order statistics $Z_{k(n),n}$ have the asymptotically normal distribution if $\min\{k(n), n-k(n)+1\} \to \infty$, as $n \to \infty$.

Exercise 10.2. Let

$$S_n = \sum_{k=1}^n d_k Z_{k,n}$$

and

$$F_k(n) = \sum_{m=k}^n \frac{d_m}{n-m+1}, \quad k = 1, 2, \ldots n.$$

Let also

$$\sum_{k=1}^n \frac{|F_k(n)|^3}{\left(\sum\limits_{k=1}^n F_k(n) \right)^3} \to 0, \tag{10.9}$$

as $n \to \infty$. Show that then random variables

$$\frac{S_n - \sum\limits_{k=1}^n F_k(n)}{\left(\sum\limits_{k=1}^n F_k^2(n) \right)^{1/2}}$$

converge in distribution to the standard normal distribution.

Exercise 10.3. Let $X_{1,n} = Z_{1,n}$ and

$$T_{k,n} = Z_{k,n} - Z_{k-1,n}, \quad k = 2, \ldots, n, \ n = 1, 2, \ldots$$

be the exponential spacings and

$$R_n = \sum_{k=1}^n b_k T_{k,n}.$$

Reformulate the result of exercise 10.2 for sums R_n.

Example 10.2. Consider now the uniform order statistics

$$0 = U_{0,n} \leqslant U_{1,n} \leqslant \cdots \leqslant U_{n,n} \leqslant U_{n+1,n} = 1, \quad n = 1, 2, \ldots.$$

From chapter 4 we know that $U_{k,n}$ has the same distribution as the ratio S_k/S_{n+1}, where

$$S_m = \sum_{k=1}^m v_k, \quad m = 1, 2, \ldots,$$

are the sums of independent $E(1)$ random variables $v_1, v_2, \ldots$. We know also that

$$EU_{k,n} = \frac{k}{n+1}.$$

Now we can see that

$$P\left\{ b_n \left(U_{k,n} - \frac{k}{n+1} \right) < x \right\} = P\left\{ b_n \left(\frac{S_k}{S_{n+1}} - \frac{k}{n+1} \right) < x \right\}$$

$$= P\left\{ \frac{b_n}{(n+1)S_{n+1}} ((n+1-k)S_k - k(S_{n+1} - S_k)) < x \right\}. \tag{10.10}$$

Due to the law of large numbers

$$\frac{S_{n+1}}{n+1} \to 1$$

in probability. Hence if

$$\frac{b_n}{(n+1)^2}\left((n+1-k)S_k - k(S_{n+1}-S_k)\right)$$

$$= \frac{b_n}{(n+1)^2}\left((n+1-k)(S_k-k) - k(S_{n+1}-S_k-(n-k+1))\right)$$

has some asymptotic distribution under the suitable choice of norming constants b_n, then this distribution is also asymptotic for

$$\frac{b_n}{(n+1)S_{n+1}}\left((n+1-k)S_k - k(S_{n+1}-S_k)\right).$$

Since random variables $(S_k - k)$ and $(S_{n+1} - S_k - (n - k + 1))$ are independent and $(S_k - k)/k^{1/2}$ and $(S_{n+1} - S_k - (n-k+1))/(n-k+1)^{1/2}$ asymptotically have the standard normal distribution if $\min\{k, n-k+1\} \to \infty$, as $n \to \infty$, we find that

$$\frac{(n+1-k)S_k - k(S_{n+1}-S_k)}{(n+1)^{1/2}k^{1/2}(n+1-k)^{1/2}}$$

also converges to the corresponding normal distribution. Moreover, if $\min\{k, n-k+1\} \to \infty$, as $n \to \infty$, and $b_n = n^{3/2}/k(n-k+1)^{1/2}$ then it provides the convergence of random variables

$$\frac{n^{3/2}\left(U_{k,n} - \frac{k}{n+1}\right)}{k^{1/2}(n-k+1)^{1/2}}$$

to the standard normal law.

Note that there are some estimates of the rate of convergence to the normal law of suitable centered and normalized uniform order staistics $U_{k,n}$ For example, the following inequality holds (see Nevzorov (2001), Chapter 8) for any $n = 1, 2, \ldots$ and $1 \leqslant k \leqslant n$:

$$\sup_x \left| P\left\{ U_{k,n} - \frac{k}{n+1} < \frac{xk^{1/2}(n-k+1)^{1/2}}{n^{3/2}} \right\} - \Phi(x) \right| \leqslant C(k^{-1/2} + (n-k+1)^{-1/2}),$$

where C is some absolute constant.

Exercise 10.4. Consider the uniform sample quantiles

$$U_{qn,n}, \quad 0 < q < 1,$$

and prove that

$$\frac{(U_{qn,n} - q)n^{1/2}}{q^{1/2}(1-q)^{1/2}}$$

converges in distribution to the standard normal distribution.

Exercise 10.5. Let

$$T_n = \sum_{k=1}^{n} c_k U_{k,n}.$$

Denote $b_{n+1} = 0$ and

$$b_m = \sum_{k=m}^{n} c_k, \quad m = 1, 2, \ldots, n.$$

Let

$$b(n) = \sum_{k=1}^{n+1} \frac{b_k}{n+1}$$

and

$$D^2(n) = \frac{1}{(n+1)^2} \sum_{k=1}^{n+1} (b_k - b(n))^2.$$

Show that if

$$L_n = \sum_{k=1}^{n+1} \frac{|b_k - b(n)|^3}{\left(\sum_{k=1}^{n+1} (b_k - b(n))^2 \right)^{3/2}} \to 0,$$

as $n \to \infty$, then random variables $(T_n - b(n))/D(n)$ have asymptotically the standard normal distribution.

Example 10.3. Consider sample quantiles $X_{qn,n}$, $0 < q < 1$, corresponding to the cdf $F(x)$ and density $f(x)$. Suppose that $f(x) > 0$ and $F(x)$ is continuous in some neighborhood of the point $G(q)$, where $G(x)$ is the inverse of F. From exercise 10.4 we know for the uniform sample quantiles $U_{qn,n}$ that

$$\frac{(U_{qn,n} - q)n^{1/2}}{q^{1/2}(1-q)^{1/2}}$$

converges to the standard normal distribution. We can try express sample quantile $X_{qn,n}$ via the corresponding uniform sample quantile as follows:

$$X_{qn,n} = G(U_{qn,n}) = G(q) + (U_{qn,n} - q)G'(q) + R_n, \tag{10.11}$$

where

$$R_n = o(|U_{qn,n} - q|)$$

and

$$P\{|R_n| > \varepsilon\} \leqslant \frac{E(U_{qn,n} - q)^2}{\varepsilon^2} \leqslant \frac{q(1-q)}{n\varepsilon^2}. \tag{10.12}$$

We can see from (10.11) and (10.12) that $X_{qn,n}$ has the same asymptotics as

$$G(q) + (U_{qn,n} - q)G'(q).$$

Hence,

$$\frac{(X_{qn,n} - G(q))\sqrt{\dfrac{n}{q(1-q)}}}{G'(q)}$$

has the same limit distribution as

$$\frac{(U_{qn,n} - q)n^{1/2}}{q^{1/2}(1-q)^{1/2}},$$

that is

$$\frac{(X_{qn,n} - G(q))\sqrt{\dfrac{n}{q(1-q)}}}{G'(q)}$$

converges to the standard normal distribution. We can also find that

$$\frac{1}{G'(q)} = f(G(q))$$

and then write that

$$(X_{qn,n} - G(q))f(G(q))\sqrt{\frac{n}{q(1-q)}}$$

asymptotically has the standard normal distribution. We can express this result as

$$X_{qn,n} \sim N\left(G(q), \frac{q(1-q)}{nf^2(G(q))}\right), \tag{10.13}$$

which means that a sequence $X_{qn,n}$ behaves asymptotically as a sequence of normal random variables with expectation $G(q)$ and variance $q(1-q)/nf^2(G(q))$.

Exercise 10.6. Consider sample medians $X_{k+1,2k+1}$ for the normal distribution with density function

$$f(x) = \frac{1}{\sqrt{2\pi}} \exp\left(-\frac{x^2}{2}\right)$$

and for the Laplace distribution with density function

$$f(x) = \frac{1}{2}\exp(-|x|)$$

and investigate the corresponding limit distributions.

Check your solutions

Exercise 10.1 (solution). As we know, sample quantiles of order p are defined as $X_{[pn]+1,n}$. It suffices to see that

$$0 < \lim_{n \to \infty} \frac{[pn]+1}{n} = p < 1$$

for any $0 < p < 1$.

Exercise 10.2 (hint). As in example 10.1 use the representation for the exponential order statistics to express S_n as the sum

$$\sum_{k=1}^{n} F_k(n) v_k,$$

where $v_1, v_2, \ldots$ is a sequence of independent exponential $E(1)$ random variables. Now it suffices to consider the corresponding Lyapunov ratio L_n and to show that condition (10.9) provides the convergence L_n to zero.

Exercise 10.3 (hint). Express R_n as

$$\sum_{k=1}^{n} d_k Z_{k,n},$$

where $d_k = b_k - b_{k+1}$, $k = 1, 2, \ldots, n$, and $d_n = b_n$.

Exercise 10.4 (hint). Use example 10.2 for $k = qn$.

Exercise 10.5 (solution). Applying the representation of the uniform order statistics $U_{k,n}$ as the ratio S_k/S_{n+1}, where

$$S_m = \sum_{k=1}^{m} v_k,$$

one gets that

$$T_n - ET_n = T_n - \sum_{k=1}^{n} c_k E \frac{S_k}{S_{n+1}} = T_n - \sum_{k=1}^{n} \frac{kc_k}{n+1} = T_n - b(n)$$

$$= \sum_{k=1}^{n} c_k U_{k,n} - b(n) \overset{d}{=} \frac{1}{S_{n+1}} \sum_{k=1}^{n+1} (b_k - b(n)) v_k.$$

It is easy to see that $S_{n+1}/(n+1)$ converges to one in distribution. Hence to show that $(T_n - b(n))/D(n)$ asymptotically has the normal distribution it suffices to prove the asymptotical normality of the sums

$$W_n = \frac{1}{(n+1)D(n)} \sum_{k=1}^{n+1} (b_k - b(n)) v_k$$

of independent random variables. We see that

$$EW_n = 0, \operatorname{Var} W_n = \sum_{k=1}^{n+1} \frac{(b_k - b(n))^2}{(n+1)^2 D^2(n)} = 1,$$

and the Lyapunov ratio for the sum W_n coincides with γL_n, where $\gamma = E|v-1|^3 = 12/e - 2$ is the third absolute central moment for the exponential distribution and

$$L_n = \sum_{k=1}^{n+1} \frac{|b_k - b(n)|^3}{\left(\sum_{k=1}^{n+1}(b_k - b(n))^2\right)^{3/2}}.$$

Hence, if L_n converges to zero as n tends to infinity, then $(T_n - b(n))/D(n)$ converges to the standard normal distribution.

Exercise 10.6 (answers). In both cases $q = 1/2$, $Q(q) = Q(1/2) = 0$. We have

$$f(Q(q)) = \frac{1}{\sqrt{2\pi}}$$

for the normal distribution and

$$f(Q(q)) = \frac{1}{2}$$

for the Laplace distribution. Using notation (10.13) we have

$$X_{k+1,2k+1} \sim N\left(0, \frac{\pi}{2(2k+1)}\right)$$

for the normal distribution and

$$X_{k+1,2k+1} \sim \overset{*}{N}\left(0, \frac{1}{2k+1}\right)$$

for the Laplace distribution.

Chapter 11

Asymptotic Behavior of the Extreme Order Statistics

We continue to study the asymptotic properties of order statistics. Below the asymptotic behavior of extreme order statistics are under investigation.

Order statistics $X_{k(n),n}$ are said to be extreme if $k = k(n)$ or $n - k(n) + 1$ is fixed, as $n \to \infty$. The most popular are maximal order statistics $X_{n,n}$ and minimal order statistics $X_{1,n}$. Both these situations correspond to $X_{n-k(n)+1}$ and $X_{k(n),n}$ for the case $k = k(n) = 1$.

Exercise 11.1. Check that if $X = -Y$, then the following relation holds for extremes $X_{n,n} = \max\{X_1, X_2, \ldots, X_n\}$ and $Y_{1,n} = \min\{Y_1, Y_2, \ldots, Y_n\}$:

$$X_{n,n} \stackrel{d}{=} -Y_{1,n}.$$

Remark 11.1. Indeed, the assertion given in exercise 11.1 can be generalized as follows:

$$\text{if } X \stackrel{d}{=} -Y,$$

$$\text{then } X_{n-k+1,n} \stackrel{d}{=} -Y_{k,n}, \text{ for any } k = 1, 2, \ldots, n.$$

Exercise 11.2. Let $F(x)$ and $f(x)$ be the distribution function and density function of X correspondingly. Find the joint distribution function $F_n(x,y)$ and joint density $f_n(x,y)$ of $X_{1,n}$ and $X_{n,n}$, $n = 2, 3, \ldots$.

Very often we need to know asymptotic distributions of $X_{1,n}$ and $X_{n,n}$, as $n \to \infty$.

Example 11.1. Due to the mentioned above relationship between maximal and minimal order statistics we can study one type of them, say, maximal ones. Consider a sequence of order statistics $M(n) = X_{n,n}$, $n = 1, 2, \ldots$. Let $F(x)$ be the distribution function of X and $\beta = \sup\{x : F(x) < 1\}$ be the right end point of the support of X. If $\beta = \infty$, then for any finite x one gets that $F(x) < 1$ and hence

$$P\{M(n) \leqslant x\} = (F(x))^n \to 0, \text{ as } n \to \infty.$$

M. Ahsanullah et al., *An Introduction to Order Statistics*,
Atlantis Studies in Probability and Statistics 3,
DOI: 10.2991/978-94-91216-83-1_11, © Atlantis Press 2013

It means that $M(n)$ converges in probability to infinity. In the case, when $\beta < \infty$, we need to distinguish two situations. If

$$P\{X = \beta\} = p > 0,$$

then

$$P\{M(n) = \beta\} = 1 - P\{M(n) < \beta\} = 1 - P^n(X < \beta) = 1 - (1 - p)^n \qquad (11.1)$$

and

$$P\{M(n) = \beta\} \to 1, \quad \text{as } n \to \infty.$$

If $P\{X = \beta\} = 0$, then we get that $P\{M(n) < \beta\} = 1$ for any n and $M(n) \to \beta$ in distribution.

Thus, we see that in all situations $M(n) \to \beta$ in distribution. This result can be sharpened if we consider the asymptotic distributions of the centered and normalized order statistics $X_{n,n}$. Indeed, if $\beta < \infty$ and $P\{X = \beta\} > 0$, then relation (11.1) gives completed information on $M(n)$ and in this case any centering and norming can not improve our knowledge about the asymptotic behavior of $M(n)$. It is clear also that if $M(n) \to \beta < \infty$, then we can choose such norming constants $b(n) \to \infty$, that $M(n)/b(n)$ converges to zero. The similar situation is also valid for the case $\beta = \infty$. In fact, if $\sup\{x : F(x) < 1\} = \infty$, we can take such sequence d_n that

$$P\{X > d_n\} = 1 - F(d_n) < \frac{1}{n^2}.$$

Indeed, $d_n \to \infty$ as $n \to \infty$. Then for any $\varepsilon > 0$ and for sufficiently large n, one gets that $\varepsilon d_n > 1$, and

$$P\left\{\frac{M(n)}{(d_n)^2} > \varepsilon\right\} \leqslant P\{M(n) > d_n\} = 1 - P\{M(n) \leqslant d_n\}$$

$$= 1 - F^n(d_n) \leqslant 1 - \left(1 - \frac{1}{n^2}\right)^n \to 0,$$

as $n \to \infty$. It means that in this case there exists also a sequence $b(n)$, say $b(n) = d_n^2$, such that $M(n)/b(n) \to 0$ in probability (and hence, in distribution). Thus, we see that the appropriate constants can provide the convergence of the normalized maxima $M(n)/b(n)$ to zero. Now the question arises, if it is possible to find centering constants $a(n)$, for which the sequence $(M(n) - a(n))$ converges to some degenerate distribution. Indeed, if $\beta < \infty$, then, as we know, $M(n) \to \beta$, and hence any centering of the type $M(n) - a(n)$, where $a(n) \to a$ (and, in particular $a(n) = 0$, $n = 1, 2, \ldots$) gives the degenerate limit distribution for the sequence $M(n) - a(n)$. Let us consider now the case, when $\beta = \infty$. We can show that in this case no centering (without norming) can provide the convergence $M(n) - a(n)$ to some finite constant. To illustrate this assertion one can take the standard exponential distribution with cdf $F(x) = 1 - \exp(-x), x > 0$.

In this situation

$$P\{M(n) - a_n < x\} = F^n(x + a_n) = (1 - \exp(-x - a(n)))^n.$$

Note that if $a_n = \log n, n \to \infty$, then

$$P\{M(n) - \log n < x\} = \left(1 - \frac{1}{n} \exp(-x)\right)^n \to \exp(-\exp(-x)) \qquad (11.2)$$

and the RHS of (11.2) presents non-degenerate distribution. Let a random variable η have the cdf

$$H_0(x) = \exp(-\exp(-x)).$$

Then we can rewrite (11.2) in the form

$$M(n) - \log n \xrightarrow{d} \eta. \qquad (11.3)$$

If we choose other centering constants a_n, then

$$M(n) - a_n = M(n) - \log n - (a_n - \log n)$$

and we see that $M(n) - a_n$ has a limit distribution only if the sequence $a_n - \log n$ converges to some finite constant a. Then

$$M(n) - a_n \xrightarrow{d} \eta - a. \qquad (11.4)$$

This means that limit distribution functions of centering maxima of independent exponentially distributed random variables $M(n) - a_n$ must have the form

$$H_a(x) = \exp(-\exp(-x - a))$$

and these distributions are nondegenerate for any a.

A new problem arises: if there exist any centering (a_n) and norming (b_n) constants, such that the sequence $(M(n) - a_n)/b_n$ converges to some nondegenerate distribution?

Exercise 11.3. Consider the exponential distribution from example 11.2. Let it be known that

$$\frac{M(n) - a_n}{b_n}, \quad b_n > 0,$$

converges to some nondegenerate distribution with a d.f. $G(x)$. Show that then

$$b_n \to b, \quad a_n - \log n \to a,$$

as $n \to \infty$, where $b > 0$ and a are some finite constants, and $G(x) = H_0(a + xb)$ with

$$H_0(x) = \exp(-\exp(-x)).$$

Exercise 11.4. Let X have the uniform $U([-1,0])$ distribution. Find the limit distribution of $nM(n)$.

Exercise 11.5. Consider X with d.f.

$$F(x) = 1 - (-x)^\alpha, \quad -1 < x < 0, \ \alpha > 0,$$

and prove that the asymptotic distribution of $n^{1/\alpha}M(n)$ has the form

$$H_{1,\alpha}(x) = \begin{cases} \exp(-(-x)^\alpha), & -\infty < x \leqslant 0, \\ 1, & x > 0. \end{cases} \tag{11.5}$$

Exercise 11.6. Let X have Pareto distribution with d.f.

$$F(x) = 1 - x^{-\alpha}, \quad x > 1.$$

Prove that the asymptotic d.f. of $M(n)/n^{1/\alpha}$ is of the form:

$$H_{2,\alpha}(x) = \begin{cases} 0, & x < 0, \\ \exp\{-x^{-\alpha}\}, & x \geqslant 0. \end{cases} \tag{11.6}$$

Remark 11.2. Changing suitably the normalized constants a_n and b_n for maximal values considered in exercises 11.5, one gets that any d.f. of the form $H_{1,\alpha}(a + bx)$, where $b > 0$ and a are arbitrary constants, can serve as the limit distribution for $(M(n) - a_n)/b_n$. Analogously, making some changing in exercise 11.6 we can prove that any d.f. of the form $H_{2,\alpha}(a + bx)$ also belongs to the set of the limit distributions for the suitably normalized maximal values.

These facts are based on the following useful result, which is due to Khinchine.

Lemma 11.1. *Let a sequence of d.f.'s F_n converge weakly to a nondegenerate d.f. G, i.e.*

$$F_n(x) \to G(x), \quad n \to \infty,$$

for any continuity point of G. Then the sequence of d.f.'s

$$H_n(x) = F_n(b_n x + a_n)$$

converges, for some constants a_n and $b_n > 0$, to a non-degenerate d.f. H if and only if $a_n \to a$, $b_n \to b$, as $n \to \infty$, where $b > 0$ and a are some constants, and $H(x) = G(bx + a)$.

Lemma 11.1 can be rewritten in the following equivalent form.

Lemma 11.2. *Let F_n be a sequence of d.f.'s and let there exist sequences of constants $b_n > 0$ and a_n such that*

$$H_n(x) = F_n(a_n + b_n x)$$

weakly converge to a non-degenerate d.f. $G(x)$. Then for some sequences of constants $\beta_n > 0$ and α_n, d.f.'s

$$R_n(x) = F_n(\alpha_n + \beta_n x)$$

weakly converge to a non-degenerate d.f. $G^*(x)$ if and only if for some $b > 0$ and a

$$\lim_{n \to \infty} \frac{\beta_n}{b_n} = b \ \text{ and } \ \lim_{n \to \infty} \frac{\alpha_n - a_n}{b_n} = a.$$

Moreover, in this situation

$$G^*(x) = G(bx + a).$$

As a corollary, we obtain from lemma 11.2 that if a sequence of maximal values $(M(n) - a_n)/b_n$ has some non-degenerate limit distribution then a sequence of $(M(n) - \alpha_n)/\beta_n$ has the same distribution if and only if

$$\beta_n \sim b_n \ \text{ and } \ \frac{\alpha_n - a_n}{b_n} \to 0, \ \text{ as } n \to \infty.$$

Considering two d.f.'s, $G(d + cx)$ and $G(a + bx)$, where $b > 0$ and $c > 0$, we say that these d.f.'s belong to the same type of distributions. Any distribution of the given type can be obtained from other distribution of the same type by some linear transformation.

Usually one of distributions, say $G(x)$, having the most simplest (or convenient) form, is chosen to represent all the distributions of the given type, which we call then G-type. As basic for their own types, we suggested above the following distributions:

$$H_0(x) = \exp(-\exp(-x));$$

$$H_{1,\alpha}(x) = \begin{cases} \exp(-(-x)^\alpha), & -\infty < x \leqslant 0, \\ 1, & x > 0; \end{cases}$$

$$H_{2,\alpha}(x) = \begin{cases} 0, & x < 0, \\ \exp\{-x^{-\alpha}\}, & x \geqslant 0, \end{cases}$$

where $\alpha > 0$.

Very often one can find that the types of distributions based on $H_0(x)$, $H_{1,\alpha}(x)$ and $H_{2,\alpha}(x)$ are named correspondingly as Gumbel, Frechet and Weibull types of limiting extreme value distributions.

Note also that any two of d.f.'s $H_{1,\alpha}$ and $H_{1,\beta}$, $\alpha \neq \beta$, do not belong to the same type, as well as d.f.'s $H_{2,\alpha}$ and $H_{2,\beta}$, $\alpha \neq \beta$. It is surprising that we can not obtain some new non-degenerate distributions, besides of $H_0(x)$-, $H_{1,\alpha}(x)$- and $H_{2,\alpha}(x)$-types, which would be limit for the suitably centering and norming maximal values.

Remark 11.3. We mentioned above that the set of all possible limit distributions for maximal values includes d.f.'s H_0, $H_{1,\alpha}$ and $H_{2,\alpha}$ only. Hence it is important for a statistician to be able to determine what d.f.'s F belong to the domains of attraction $D(H_0)$, $D(H_{1,\alpha})$ and $D(H_{2,\alpha})$ of the corresponding limit laws.

We write that $F \in D(H)$, if the suitably normalized maximal values $M(n)$, based on X's with a common d.f. F, have the limit d.f. H. For instance, if

$$F(x) = 1 - \exp(-x), \quad x > 0,$$

then $F \in D(H_0)$. If X's are the uniformly $U([a,b])$ distributed random variables with

$$F(x) = \frac{x-a}{b-a}, \quad a < x < b,$$

then $F \in D(H_{1,1})$. If

$$F(x) = 1 - x^{-\alpha}, \quad x > 1 \text{ (Pareto distribution)},$$

then $F \in D(H_{2,\alpha})$. There are necessary and sufficient conditions for f to belong $D(H_0)$, $D(H_{1,\alpha})$ and $D(H_{2,\alpha})$ but their form is rather cumbersome.

As an example of the corresponding results one can find below the following theorem which is due to Gnedenko (Gnedenko, 1943).

Theorem 11.1. *Let* $X_1, X_2, \ldots$ *be a sequence of independent identically distributed random variables with distribution function* $F(x)$ *and let* $b = \sup\{x : F(x) < 1\}$. *Then* F *belongs* $D(H_0)$, *if and only if there exists a positive function* $g(t)$ *such that*

$$\frac{(1 - F(t + xg(t)))}{1 - F(t)} = \exp(-x),$$

for all real x.

The analogous conditions were obtained for f *which belong* $D(H_{1,\alpha})$ *and* $D(H_{2,\alpha})$.

Hence, simple sufficient conditions are more interesting for us. We present below some of them.

Theorem 11.2. *Let d.f.* F *have positive derivative* F' *for all* $x > x_0$. *If the following relation is valid for some* $\alpha > 0$:

$$\frac{xF'(x)}{1 - F(x)} \to \alpha, \tag{11.7}$$

as $x \to \infty$, *then* $F \in D(H_{2,\alpha})$. *The centering,* a_n, *and normalizing,* b_n, *constants can be taken to satisfy relations*

$$a_n = 0 \quad \text{and} \quad F(b_n) = 1 - \frac{1}{n}. \tag{11.8}$$

Theorem 11.3. *Let d.f. F have positive derivative F' for x in some interval (x_1, x_0) and $F'(x) = 0$ for $x > x_0$. If*

$$\frac{(x - x_0)F'(x)}{1 - F(x)} \to \alpha, \quad x \to x_0, \tag{11.9}$$

then

$$F \in D(H_{1,\alpha}).$$

The centering, a_n, and normalizing, b_n, constants can be taken to satisfy relations

$$a_n = x_0 \text{ and } F(x_0 - b_n) = 1 - \frac{1}{n}.$$

Theorem 11.4. *Let d.f. F have negative second derivative $F''(x)$ for x in some interval (x_1, x_0), and let $F'(x) = 0$ for $x > x_0$. If*

$$\frac{F''(x)(1 - F(x))}{(F'(x))^2} = -1,$$

then

$$F \in D(H_0).$$

The centering, a_n, and normalizing, b_n, constants can be taken to satisfy relations

$$F(a_n) = 1 - \frac{1}{n} \text{ and } b_n = h(a_n),$$

where

$$h(x) = \frac{1 - F(x)}{F'(x)}.$$

Exercise 11.7. Let

$$F(x) = \frac{1}{2} + \frac{1}{\pi} \arctan x \quad \text{(the Cauchy distribution)}.$$

Prove that

$$F \in D(H_{2,\alpha}).$$

What normalizing constants, a_n and b_n, can be taken in this case?

Exercise 11.8. Let

$$F(x) = \frac{1}{\sqrt{2\pi}} \int_{-\infty}^{x} \exp\left(-\frac{t^2}{2}\right) dt.$$

Show that

$$F \in D(H_0)$$

and the normalizing constants can be taken as follows:

$$a_n = (2\log n - \log\log n - \log 4\pi)^{1/2} \text{ and } b_n = \frac{1}{a_n}.$$

Exercise 11.9. Use theorem 11.3 to find the limit distribution and the corresponding normalizing constants for gamma distribution with p.d.f.

$$f(x) = \frac{x^{\alpha-1}\exp(-x)}{\Gamma(\alpha)}, \quad x > 0, \ \alpha > 0.$$

Example 11.2. Above we considered the most popular distributions (exponential, uniform, normal, Cauchy, gamma, Pareto) and found the correspondence between these distributions and limit distributions for maximal values. It is interesting to mention that there are distributions, which are not in the domain of attraction of any limit law, H_0, $H_{1,\alpha}$ or $H_{2,\alpha}$. In fact, let $X_1, X_2, \ldots$ have geometric distribution with probabilities

$$p_n = P\{X = n\} = (1-p)p^n, \quad n = 0, 1, 2, \ldots.$$

Then

$$F(x) = P\{X \leqslant x\} = 1 - p^{[x+1]}, \quad x \geqslant 0,$$

where $[x]$ denotes the entire part of x. Suppose that $H(x)$ (one of functions H_0, $H_{1,\alpha}$, $H_{2,\alpha}$) is a limit d.f. for $(M(n) - a_n)/b_n$ under the suitable choice of normalizing constants a_n and b_n. It means that

$$F^n(a_n + xb_n) \to H(x), \quad n \to \infty,$$

or

$$n\log(1 - (1 - F(a_n + xb_n)) \to \log H(x). \tag{11.10}$$

Due to the relation

$$\log(1 - x) \sim -x, \quad x \to 0,$$

we get from (11.10) that

$$n(1 - F(a_n + xb_n)) \to -\log H(x),$$

and hence

$$\log n + \log(1 - F(a_n + xb_n)) \to h(x) = \log(-\log H(x)), \tag{11.11}$$

as $n \to \infty$, for all x such that $0 < H(x) < 1$. In our case

$$\log(1 - F(a_n + xb_n)) = [1 + a_n + xb_n]\log p$$

and (11.11) can be rewritten as

$$\log n + [1 + a_n + xb_n]\log p \to h(x).$$

One can express $[1 + a_n + xb_n]$ as $a_n + \gamma_n + xb_n$, where $0 = \gamma_n < 1$. Then

$$(\log n + (a_n + \gamma_n) \log p) + xb_n \log p \to h(x). \tag{11.12}$$

It follows from (11.12) that $b_n \to b$, where b is a positive constant. We get from lemma 11.2 that if

$$F^n(a_n + xb_n) \to H(x),$$

then

$$F^n(a_n + xb) \to H(x). \tag{11.13}$$

Fix some x and take $x_1 = x - 1/3b$ and $x_2 = x + 1/3b$. The difference between $a_n + x_2b$ and $a_n + x_1b$ is less than one. It means that for any n, at least two of three points $[a_n + x_1b]$, $[a_n + xb]$ and $[a_n + x_2b]$ must coincide. Then sequences

$$F^n(a_n + x_1b) = (1 - p^{[a_n + bx_1]})^n,$$
$$F^n(a_n + xb) = (1 - p^{[a_n + bx]})^n,$$

and

$$F^n(a_n + x_2b) = (1 - p^{[a_n + bx_2]})^n$$

can not have three different limits $H(x_1) < H(x) < H(X_2)$. This contradicts to proposal that $(M(n) - a_n)/b_n$ has a non-degenerate limit distribution.

Returning to results of exercise 11.1 one can find the possible types of limit distributions for minimal values

$$m(n) = \min\{X_1, \ldots, X_n\}.$$

It appears that the corresponding set of non-degenerate asymptotic d.f.'s for the suitably normalized minimal values are defined by the following basic d.f.'s:

$$L_0(x) = 1 - \exp(-\exp(x));$$

$$L_{1,\alpha}(x) = \begin{cases} 0, & x < 0, \\ 1 - \exp(-x^\alpha), & 0 \leqslant x < \infty; \end{cases}$$

$$L_{2,\alpha}(x) = \begin{cases} 1 - \exp\{-(-x)^{-\alpha}\}, & x < 0, \\ 1, & x \geqslant 0, \end{cases}$$

where $\alpha > 0$.

Above we considered the situation with the asymptotic behavior of extremes $X_{n,n}$ and $X_{1,n}$. Analogous methods can be applied to investigate the possible asymptotic distributions of the k-th extremes - order statistics $X_{n-k+1,n}$ and $X_{k,n}$, when $k = 2, 3, \ldots$ is some fixed number and n tends to infinity. The following results are valid in these situations.

Theorem 11.5. *Let random variables $X_1, X_2, \ldots$ be independent and have a common d.f. F and $X_{n-k+1,n}$, $n = k, k+1, \ldots$, be the $(n-k+1)$-th order statistics. If for some normalizing constants a_n and b_n,*

$$P\left\{\frac{X_{n,n} - a_n}{b_n} < x\right\} \to T(x)$$

in distribution, as $n \to \infty$, then the limit relation

$$P\left\{\frac{X_{n-k+1,n} - a_n}{b_n} < x\right\} \to T(x) \sum_{j=0}^{k-1} \frac{(-\log T(x))^j}{j!} \qquad (11.14)$$

holds for any x, as $n \to \infty$.

Theorem 11.6. *Let random variables $X_1, X_2, \ldots$ be independent and have a common d.f. F and $X_{k,n}$, $n = k, k+1, \ldots$, be the k-th order statistics. If for some normalizing constants a_n and b_n,*

$$P\left\{\frac{X_{1,n} - a_n}{b_n} < x\right\} \to H(x)$$

in distribution, then the limit relation

$$P\left\{\frac{X_{k,n} - a_n}{b_n} < x\right\} \to H(x) \sum_{j=0}^{k-1} \frac{(-\log H(x))^j}{j!} \qquad (11.15)$$

is valid for any x, as $n \to \infty$.

Exercise 11.10. Let $F(x) = 1 - \exp(-x)$, $x > 0$. Find the corresponding limit (as $n \to \infty$) distributions, when $k = 2, 3, \ldots$ is fixed, for sequences

$$Y_n = (X_{n-k+1,n} - \log n),$$

Check your solutions

Exercise 11.1 (solution). Let $F(x) = P\{X \leqslant x\}$. We must compare

$$P\{X_{n,n} \leqslant x\} = F^n(x)$$

and

$$P\{-Y_{1,n} \leqslant x\} = P\{Y_{1,n} \geqslant -x\} = P^n\{Y \geqslant -x\} = P^n\{-Y \leqslant x\} = P^n\{X \leqslant x\} = F^n(x).$$

Thus, we see that

$$X_{n,n} \overset{d}{=} -Y_{1,n}.$$

Exercise 11.2 (solution). It is clear that

$$F_n(x,y) = F^n(y),$$

if $x \geqslant y$. Let $x < y$. Then

$$
\begin{aligned}
F_n(x,y) &= P\{X_{1,n} \leqslant x, \, X_{n,n} \leqslant y\} \\
&= P\{X_{n,n} \leqslant y\} - P\{X_{1,n} > x, \, X_{n,n} \leqslant y\} \\
&= F^n(y) - P^n\{x < X \leqslant y\} \\
&= F^n(y) - (F(y) - F(x))^n.
\end{aligned}
$$

Now it is evident, that

$$
f_n(x,y) = \begin{cases} 0, & \text{if } x \geqslant y, \\ n(n-1)(F(y) - F(x))^{n-2} f(y) f(x), & \text{if } x < y. \end{cases}
$$

Exercise 11.3 (solution). From (11.3) we know that for any x,

$$P\{M(n) - \log n \leqslant x\} \to H_0(x), \quad n \to \infty. \tag{11.16}$$

Moreover, since $H_0(x)$ is a continuous d.f., it follows from (11.16) that

$$\sup_x |P\{M(n) - \log n \leqslant x\} - H_0(x)| \to 0. \tag{11.17}$$

From condition of exercise 11.3 we have also that

$$P\left\{ \frac{M(n) - a_n}{b_n} \leqslant x \right\} \to G(x), \quad n \to \infty, \tag{11.18}$$

where $G(x)$ is some non-degenerate d.f.

One sees that

$$P\left\{ \frac{M(n) - a_n}{b_n} \leqslant x \right\} - G(x) =$$

$$(P\{(M(n) - \log n) \leqslant x b_n + (a_n - \log n)\} - H_0(x b_n + (a_n - \log n))) +$$

$$(G(x) - H_0(x b_n + (a_n - \log n))) = I_1(x) + I_2(x) \to 0. \tag{11.19}$$

Due to (11.17), $I_1(x) \to 0$, and hence (11.19) implies that

$$H_0(x b_n + (a_n - \log n)) \to G(x). \tag{11.20}$$

In particular,

$$H_0(a_n - \log n) \to G(0),$$

as $n \to \infty$. It means that a sequence $a_n - \log n$ has a limit, say, a.

Suppose that $G(0) = 0$. Then $a = -\infty$. Since G is a non-degenerate d.f., there is a finite positive c, such that $0 < G(c) < 1$. It means that

$$cb_n + (a_n - \log n) \to \gamma = H_0^{-1}(G(c)), \qquad (11.21)$$

where

$$H_0^{-1}(x) = -\log(-\log x)$$

is the inverse function of H_0. Relation (11.21) implies that $b(n)$ tends to infinity, as $n \to \infty$, and

$$xb_n + (a_n - \log n) = (x - c)b_n + (cb_n + a_n - \log n) \to \infty$$

for any $x > c$. In this case $G(x) = 1$ for $x > c$ and hence $G(c) = 1$. This contradicts to our statement that $G(c) < 1$. Hence $G(0) > 0$. Analogously we can prove that $G(0) < 1$. Thus,

$$\lim_{n \to \infty} (a_n - \log n) = a, \quad -\infty < a < \infty.$$

Now we get from (11.21) that the sequence $b(n)$ also has the finite limit $b = (\gamma - a)/c$. Taking into account relation (11.20) and increase of $G(x)$, we obtain that $b > 0$. It is evident also that

$$G(x) = \lim_{n \to \infty} H_0(xb(n) + (a_n - \log n)) = H_0(a + xb).$$

Exercise 11.4 (solution). In this case, according to example 11.1, $M(n)$ itself has the degenerate distribution, concentrated at zero, while normalized maxima has non-degenerate distribution. In fact, we have

$$F(x) = 1 + x, \quad -1 < x < 0,$$

and for any $x \leqslant 0$,

$$P\{nM(n) \leqslant x\} = \left(F\left(\frac{x}{n}\right)\right)^n = \left(1 + \frac{x}{n}\right)^n \to \exp(x),$$

as $n \to \infty$. Indeed, $nM(n) \leqslant 0$ and hence,

$$P\{nM(n) \leqslant x\} = 1$$

for any n and $x > 0$.

Exercise 11.5 (solution). In this situation we simply generalize the case given in exercise 11.4:

$$P\{n^{1/\alpha}M(n) \leqslant x\} = (F(x/n^{1/\alpha}))^n = \left(1 - \frac{(-x)^\alpha}{n}\right)^n \to \exp(-x^\alpha), \quad -\infty < x \leqslant 0.$$

Exercise 11.6 (solution). We have

$$P\{n^{-1/\alpha}M(n) \leqslant x\} = \left(F(xn^{1/\alpha})\right)^n = \left(1 - \frac{x^{-\alpha}}{n}\right)^n \to \exp(-x^{-\alpha}), \quad x \geqslant 0.$$

Exercise 11.7 (hint and answer). Use relations

$$(1 - F(x)) \sim \frac{1}{\pi x},$$

$$F'(x) \sim \frac{1}{\pi x^2}, \quad x \to \infty,$$

and the statement of theorem 11.1 to show that it is possible to take $a_n = 0$ and $b_n = n$.

Exercise 11.8 (hint). Show that

$$1 - F(x) \sim \frac{F'(x)}{x} = \frac{1}{x\sqrt{2\pi}} \exp\left(-\frac{x^2}{2}\right),$$

$$F''(x) \sim -xF'(x), \quad x \to \infty$$

and then use the statement of theorem 11.3. To find constants a_n you need to solve the equation

$$\frac{1}{n} = 1 - F(a_n) \sim \frac{1}{a_n\sqrt{2\pi}} \exp\left(-\frac{(a_n)^2}{2}\right)$$

or simply to check that the sequence

$$a_n = (2\log n - \log\log n - \log 4\pi)^{1/2}$$

satisfies this equation. Theorem 11.3 recommends to take $b_n = h(a_n)$, where

$$h(x) = 1 - \frac{F(x)}{F'(x)},$$

but we can see that

$$h(x) \sim \frac{1}{x}, \quad x \to \infty,$$

and hence

$$h(a_n) \sim \frac{1}{a_n}, \quad n \to \infty.$$

From lemma 11.2 we know that the sequence of constants b_n can be changed by a sequence of equivalent constants and this changing saves the limit distribution for maximal values. Moreover, going further we can take the following more simple sequence of equivalent norming constants, namely $b_n = (2\log n)^{1/2}$.

Exercise 11.9 (answer). In this case $F \in D(H_0)$, $a_n = \log n$ and $b_n = 1$.

Exercise 11.10 (hint and answer). Use relation (11.2) and theorem 11.5 to find that

$$P\{Y_n < x\} \to \exp(-e^{-x}) \sum_{j=0}^{k-1} \frac{e^{-jx} \exp(-e^{-x})}{j!}, \quad \text{as } n \to \infty.$$

Below we give the table 11.11 in which the limit distributions of extreme $X_{n.n}$, the corresponding centering and normalizing constants are given for the most popular distributions. Here we denote three types of the possible asymptotic distributions with their respective distribution function as:

$$T_{10} : \quad F(x) = \exp(-e^{-x}), \qquad -\infty < x < \infty$$

$$T_{2\alpha} : \quad F(x) = \exp(-x^{-\alpha}), \qquad x > 0, \ \alpha < 0$$

$$T_{3\alpha} : \quad F(x) = \exp(-(-x)^{\alpha}), \qquad x < 0, \ \alpha > 0.$$

Table 11.1: Normalizing Constants for $X_{n,n}$

Distribution	$f(x)$	a_n	b_n	Domain
Beta	$cx^{\alpha-1}(1-x)^{\beta-1}$, $c = \dfrac{\Gamma(\alpha+\beta)}{\Gamma(\alpha)\Gamma(\beta)}$, $\alpha > 0,\ \beta > 0,\ 0 < x < 1$	1	$\left(\dfrac{\beta}{nc}\right)^{1/\beta}$	$T_{3\beta}$
Cauchy	$\dfrac{1}{\pi(1+x^2)}$, $-\infty < x < \infty$	0	$\cotan\left(\dfrac{\pi}{n}\right)$	T_{21}
Discrete Pareto	$P(X = k) = [k]^\theta$ $-[k+1]^\theta,\ \theta > 0, k \geqslant 1, [\ ]$ represents the greatest integer contained in	0	$n^{1/\theta}$	$T_{2/\theta}$
Exponential	$\sigma e^{-\sigma x}, 0 < x < \infty, \sigma > 0$	$\dfrac{1}{\sigma}\ln n$	$\dfrac{1}{\sigma}$	T_{10}
Gamma	$\dfrac{x^{\alpha-1}e^{-x}}{\Gamma(\alpha)}, 0 < x < \infty$	$\ln n + \ln\Gamma(\alpha)$ $-(\alpha-1)\ln\ln n$	1	T_{10}
Laplace	$\dfrac{1}{2}e^{-|x|}, -\infty < x < \infty$	$\ln\left(\dfrac{n}{2}\right)$	1	T_{10}
Logistic	$\dfrac{e^{-x}}{(1+e^{-x})^2}$	$\ln n$	1	T_{10}
Lognormal	$\dfrac{1}{x\sqrt{2\pi}}e^{-\frac{1}{2}(\ln x)^2}$, $0 < x < \infty$	e^{α_n}, $\alpha_n = \dfrac{1}{\beta_n} - \dfrac{\beta_n D_n}{2}$, $D_n = \ln\ln n + \ln 4\pi$, $\beta_n = (2\ln n)^{-1/2}$	$(2\ln n)^{-1/2}e^{\alpha_n}$	T_{10}
Normal	$\dfrac{1}{\sqrt{2\pi}}e^{-\frac{1}{2}x^2}$, $-\infty < x < \infty$	$\dfrac{1}{\beta_n} - \dfrac{\beta_n D_n}{2}$, $D_n = \ln\ln n + \ln 4\pi$, $\beta_n = (2\ln n)^{-1/2}$	$(2\ln n)^{-1/2}$	T_{10}
Pareto	$\alpha x^{-(\alpha+1)}, x > 1, \alpha > 0$	0	$n^{1/\alpha}$	$T_{2\alpha}$
Power Function	$\alpha x^{\alpha-1}, 0 < x < 1, \alpha > 0$	1	$\dfrac{1}{n\alpha}$	T_{31}
Rayleigh	$\dfrac{2x}{\sigma^2}e^{-x^2/\sigma^2}, x \geqslant 0$	$\alpha(\ln n)^{1/2}$	$\dfrac{\alpha}{2}(\ln n)^{-1/2}$	T_{10}
t distribution	$\dfrac{k}{(1+x^2/v)^{(v+1)/2}}$, $k = \dfrac{\Gamma((v+1)/2)}{(\pi v)^{1/2}\Gamma(v/2)}$	0	$\left(\dfrac{kn}{v}\right)^{1/v}$	T_{2v}
Truncated Exponential	$Ce^{-x}, C = 1/(1-e^{-e(F)})$, $0 < x < e(F) < \infty$	$e(F)$	$\dfrac{e^{e(F)}-1}{n}$	T_{31}
Type 1	$e^{-x}e^{-e^{-x}}$	$\ln n$	1	T_{10}

Continued on next page

Table 11.1 – continued from previous page

Distribution	$f(x)$	a_n	b_n	Domain
Type 2	$\alpha x^{-(\alpha+1)}e^{-x^{-\alpha}}$, $\quad x > 0, \alpha > 0$	0	$n^{1/\alpha}$	$T_{2\alpha}$
Type 3	$\alpha(-x)^{\alpha-1}e^{-(-x)^{\alpha}}$, $\quad x < 0, \alpha > 0$	0	$n^{-1\alpha}$	$T_{3\alpha}$
Uniform	$1/\theta, 0 < x < \theta$	θ	θ/n	T_{31}
Weibull	$\alpha x^{\alpha-1}e^{-x^{\alpha}}, x > 0, \alpha > 0$	$(\ln n)^{1/\alpha}$	$\dfrac{(\ln n)^{(1-\alpha)/\alpha}}{\alpha}$	$T_{1\alpha}$

Chapter 12

Some Properties of Estimators Based on Order Statistics

We investigate some properties of estimators based on order statistics. Examples of sufficient statistics are given. We also consider unbiased estimators of some unknown parameters as well as estimators with minimum mean square errors.

Order statistics play a very important role in different statistical problems. Very often they are used as the best (in some sense) or simply as convenient estimators of unknown parameters. Below we will study different properties of various estimators based on order statistics.

We begin from the definition of sufficient statistics.

Definition 12.1. Suppose $X_1, X_2, \ldots, X_n$ are n observations from a distribution with cdf as $F_\theta(x)$. Here θ is an unknown parameter whose possible values form a set Θ. Let

$$T = T(X_1, X_2, \ldots, X_n).$$

be a known function of $X_1, X_2, \ldots, X_n$. Then, T is called a sufficient statistic for θ if the conditional distribution of $X_1, X_2, \ldots, X_n$ given $T = t$ does not depend on $\theta \in \Theta$.

Remark 12.1. The statistic T and parameter θ in definition 12.1 may be vectors.

Example 12.1. Let $X_1, X_2, \ldots, X_n$ be n observations from an absolutely continuous distribution with cdf as $F_\theta(x)$, $\theta \in \Theta$. Consider the vector of order statistics $T = (X_{1,n}, X_{2,n}, \ldots, X_{n,n})$. Then T is sufficient for $\theta \in \Theta$. In fact, let $f_\theta(x)$ be the pdf of $f_\theta(x)$. It is not difficult to check that

$$P\{X_1 = \beta_1, \ldots, X_n = \beta_n \mid X_{1,n} = x_1, \ldots, X_{n,n} = x_n\} = \frac{1}{n!} \tag{12.1}$$

for all n! permutations $(\beta_1, \ldots, \beta_n)$ of arbitrary chosen values $x_1, \ldots, x_n$, such that $x_1 \leqslant \cdots \leqslant x_n$ and $f_\theta(x_k) > 0$, for any $k = 1, 2, \ldots, n$ and $\theta \in \Theta$. We see that the RHS of (12.1) does not depend on θ. Thus, T is a sufficient statistic for $\theta \in \Theta$.

M. Ahsanullah et al., *An Introduction to Order Statistics*,
Atlantis Studies in Probability and Statistics 3,
DOI: 10.2991/978-94-91216-83-1_12, © Atlantis Press 2013

Remark 12.2. There exists the so-called factorization criterion to check if some statistic $T = T(X_1, X_2, \ldots, X_n)$ is sufficient for a parameter $\theta \in \Theta$. We formulate it for absolutely continuous distributions in the following form.

Theorem 12.1. *Let* $X_1, X_2, \ldots, X_n$ *be* n *independent observations from a distribution with a density function* $f_\theta(x)$, $\theta \in \Theta$. *Then*

$$T = T(X_1, X_2, \ldots, X_n)$$

is a sufficient statistic for θ, *if the joint pdf of* $X_1, X_2, \ldots, X_n$ *having the form*

$$f_\theta(x_1, x_2, \ldots, x_n) = f_\theta(x_1) f_\theta(x_2) \cdots f_\theta(x_n)$$

may be expressed as follows:

$$f_\theta(x_1, x_2, \ldots, x_n) = h(x_1, x_2, \ldots, x_n) g(T(x_1, x_2, \ldots, x_n), \theta), \qquad (12.2)$$

where h *is a nonnegative function of* $x_1, x_2, \ldots, x_n$ *only and does not depend on* θ, *and* g *is a nonnegative function of* θ *and* $T(x_1, x_2, \ldots, x_n)$ *only.*

Example 12.2. Let $X_1, X_2, \ldots, X_n$ be n observations from a uniform $U(0, \theta)$, distribution with pdf $f_\theta(x)$, which equals $1/\theta$, if $0 < x < \theta$, where $\theta > 0$ is an unknown parameter, and it is zero otherwise. We will show that

$$T(X_1, X_2, \ldots, X_n) = \max\{X_1, X_2, \ldots, X_n\} = X_{n,n}$$

is a sufficient statistic for the parameter θ.

One can write the joint probability density function of the observations $X_1, X_2, \ldots, X_n$ as follows:

$$f_\theta(x_1, x_2, \ldots, x_n) = \frac{1}{\theta^n} \prod_{i=1}^{n} R(x_i) C(x_i, \theta),$$

where

$$C(x, \theta) = \begin{cases} 1, & \text{if } x < \theta, \\ 0, & \text{otherwise,} \end{cases}$$

and

$$R(x) = \begin{cases} 0, & \text{if } x \leqslant 0, \\ 1, & \text{if } x > 0. \end{cases}$$

We note that

$$\prod_{i=1}^{n} C(x_i, \theta)$$

coincides with the function

$$C(\max(x_1,\ldots,x_n),\theta),$$

which can be written also as

$$C(T(x_1,\ldots,x_n),\theta).$$

Hence, in our case relation (12.2) holds with functions

$$h(x_1,\ldots,x_n) = \prod_{i=1}^{n} R(x_i),$$

which does not depend on θ, and

$$g(x,\theta) = \frac{C(x,\theta)}{\theta^n}.$$

Thus,

$$T(X_1,X_2,\ldots,X_n) = \max\{X_1, X_2,\ldots,X_n\} = X_{n,n}$$

is a sufficient statistic for the parameter θ.

Exercise 12.1. Let $X_1, X_2,\ldots,X_n$ be n independent observations from a uniform $U(-\theta,\theta)$ distribution with the density function $f_\theta(x)$, which equals $1/2\theta$, if $|x| < \theta$. Prove that

$$T = \max\{|X_{1,n}|, X_{n,n}\}$$

is a sufficient statistic for the parameter θ, $\theta > 0$.

Exercise 12.2. Let $X_1, X_2,\ldots,X_n$ be n observations from an exponential distribution with pdf $f_\theta(x)$, defined as follows:

$$f_\theta(x) = \begin{cases} e^{-(x-\theta)}, & \theta < x < \infty, \\ 0, & \text{otherwise.} \end{cases}$$

Show that

$$T = X_{1,n}$$

is a sufficient statistic for θ, $\theta > 0$.

The next object for investigation are unbiased estimators.

Definition 12.2. If a statistic

$$T = T(X_1,X_2,\ldots,X_n)$$

is such that $E(T) = \theta$, for any $\theta \in \Theta$, then T is called an *unbiased estimator* of θ.

Example 12.3. Suppose that $X_1, X_2, \ldots, X_n$ are n observations from a uniform distribution with pdf $f_\theta(x)$, which is defined as follows:

$$f_\theta(x) = \begin{cases} \dfrac{1}{\theta}, & 0 < x < \theta, \ \theta > 0, \\ 0, & \text{otherwise.} \end{cases}$$

Then

$$\frac{n+1}{n} X_{n,n}$$

is an unbiased estimator of θ. In fact,

$$E\left(\frac{n+1}{n} X_{n,n}\right) = \int_0^\theta \frac{n+1}{n} xn \frac{x^{n-1}}{\theta^n} dx = \theta.$$

Example 12.4. Suppose that $X_1, X_2, \ldots, X_n$ are n observations from an exponential distribution with pdf $f_\theta(x)$ given as follows:

$$f_\theta(x) = \begin{cases} \exp(-(x-\theta)), & \text{if } x > \theta, \\ 0, & \text{otherwise.} \end{cases}$$

To show that

$$T = X_{1,n} - \frac{1}{n}$$

is an unbiased estimator for θ, we use equalities

$$E(X_{1,n}) = \int_\theta^\infty xn e^{-n(x-\theta)} dx = \theta + \frac{1}{n}.$$

Thus,

$$ET = E\left(X_{1,n} - \frac{1}{n}\right) = \theta$$

and hence T is an unbiased estimator of θ.

Exercise 12.3. Suppose that $X_1, X_2, \ldots, X_n$ are n observations from a uniform distribution with pdf $f_\theta(x)$ as

$$f_\theta(x) = \begin{cases} 1, & \theta - \dfrac{1}{2} < x < \theta + \dfrac{1}{2}, \\ 0, & \text{otherwise.} \end{cases}$$

Prove that

$$T = \frac{X_{1,n} + X_{n,n}}{2}$$

is an unbiased estimator of θ.

Some statistical definitions are connected with the so-called *mean square error*. One says that an estimator

$$T^* = T^*(X_1, X_2, \ldots, X_n, \theta)$$

is at least as good as any other estimator

$$T = T(X_1, X_2, \ldots, X_n, \theta)$$

in the sense of the mean square error if

$$E(T^* - \theta)^2 \leqslant E(T - \theta)^2$$

for all $\theta \in \Theta$. If we restrict to unbiased estimators, then

$$\text{Var}(T^*) = E(T^* - \theta)^2 \leqslant E(T - \theta)^2 = \text{Var}(T)$$

and the estimator T^* is called a minimum variance unbiased estimator. In general, a particular estimator will be better than another for some values of θ and worse for other. Sometimes we need to compare mean square errors of estimators, which belong to some set of convenient statistics.

Example 12.5. Suppose that $X_1, X_2, \ldots, X_n$ are n independent observations from a uniform $U(0, \theta)$ distribution with pdf $f_\theta(x)$ given as follows:

$$f_\theta(x) = \begin{cases} \dfrac{1}{\theta}, & 0 < x < \theta, \ \theta > 0, \\ 0, & \text{otherwise.} \end{cases}$$

We consider statistics

$$T = aX_{n,n}, \quad -\infty < a < \infty,$$

and find among them an estimator of θ, having the minimum mean square error. Since

$$EX_{n,n} = \frac{n\theta}{n+1}$$

and

$$E(X_{n,n})^2 = \frac{n\theta^2}{n+2},$$

we obtain equality

$$E(aX_{n,n} - \theta)^2 = a^2 \frac{n}{n+2}\theta^2 - 2a\frac{n}{n+1}\theta^2 + \theta^2.$$

One sees that

$$E(aX_{n,n} - \theta)^2$$

is minimum if

$$a = \frac{n+2}{n+1}.$$

Thus,

$$T^* = \frac{n+2}{n+1} X_{n,n}$$

has the minimum mean square error among all estimators of θ having the form $aX_{n,n}$ and this mean square error of T^* has the value

$$E(T^* - \theta)^2 = \frac{1}{(n+1)^2} \theta^2.$$

Let us take now an unbiased estimator of θ, which has the form

$$T_1 = \frac{n+1}{n} X_{n,n}.$$

It is easy to check that

$$\text{Var}(T_1) = E(T_1 - \theta)^2 = \frac{1}{n(n+2)} \theta^2 > \frac{1}{(n+1)^2} \theta^2 = E(T^* - \theta)^2.$$

It means that the mean square error of a biased estimator of some parameter may be less than the mean square error of an unbiased estimator of the same parameter.

Check your solutions

Exercise 12.1 (hint). Use the construction, which is analogous to one given in example 12.2.

Exercise 12.2 (solution). For this case

$$f_\theta(x_1, x_2, \ldots, x_n) = \begin{cases} \exp\{-(x_1 + x_2 + \cdots + x_n) + n\theta\}, & \text{if } \min\{x_1, x_2, \ldots, x_n\} > \theta, \\ 0, & \text{otherwise.} \end{cases}$$

Hence,

$$f_\theta(x_1, x_2, \ldots, x_n) = \exp\{-(x_1 + x_2 + \cdots + x_n)\} \exp(n\theta) R(\min\{x_1, x_2, \ldots, x_n\}, \theta),$$

where

$$R(x, \theta) = \begin{cases} 1, & \text{if } x > \theta, \\ 0, & \text{otherwise.} \end{cases}$$

It means that

$$T(X_1, \ldots, X_n) = \min\{X_1, \ldots, X_n\} = X_{1,n}$$

is a sufficient statistic for the parameter θ.

Exercise 12.3 (hint). Check that

$$E(X_{1,n}) = \theta - \frac{n-1}{2(n+1)}$$

and

$$E(X_{n,n}) = \theta + \frac{n-1}{2(n+1)}.$$

Chapter 13

Minimum Variance Linear Unbiased Estimators

We give definitions and examples of the best (in the sense of minimum variance) unbiased estimators of the location and scale parameters, based on linear combinations of order statistics.

In chapter 12 we introduced the definition of a minimum variance unbiased estimator. Now we will consider some examples of such estimators, which are expressed as linear combinations of order statistics. We will call them the minimum variance linear unbiased estimators (MVLUEs). Let us begin from theMVLUEs of location and scale parameters. Suppose that X has an absolutely continuous distribution function of the form

$$F\left(\frac{x-\mu}{\sigma}\right), \quad -\infty < \mu < \infty, \ \sigma > 0.$$

Further assume

$$E(X_{r,n}) = \mu + \alpha_r \sigma, \quad r = 1, 2, \ldots, n,$$
$$\text{Var}(X_{r,n}) = V_{rr}\sigma^2, \quad r = 1, 2, \ldots, n,$$
$$\text{Cov}(X_{r,n}, X_{s,n}) = \text{Cov}(X_{s,n}, X_{r,n}) = V_{rs}\sigma^2, \quad 1 \leqslant r < s \leqslant n.$$

Let

$$\mathbf{X}' = (X_{1,n}, X_{2,n}, \ldots, X_{n,n}).$$

We can write

$$E(\mathbf{X}) = \mu \mathbf{1} + \sigma \boldsymbol{\alpha} \tag{13.1}$$

where

$$\mathbf{1} = (1, 1, \ldots, 1)',$$

$$\boldsymbol{\alpha} = (\alpha_1, \alpha_2, \ldots, \alpha_n)'$$

M. Ahsanullah et al., *An Introduction to Order Statistics*,
Atlantis Studies in Probability and Statistics 3,
DOI: 10.2991/978-94-91216-83-1_13, © Atlantis Press 2013

and

$$\text{Var}(\mathbf{X}) = \sigma^2 \boldsymbol{\Sigma},$$

where $\boldsymbol{\Sigma}$ is a matrix with elements V_{rs}, $1 \leqslant r, s \leqslant n$.

Then the MVLUE's of the location and scale parameters μ and σ are

$$\hat{\mu} = \frac{1}{\Delta} \{ \alpha' \Sigma^{-1} \alpha \, 1' \Sigma^{-1} - \alpha' \Sigma^{-1} 1 \, \alpha' \Sigma^{-1} \} X \tag{13.2}$$

and

$$\hat{\sigma} = \frac{1}{\Delta} \{ 1' \Sigma^{-1} 1 \, \alpha' \Sigma^{-1} - 1' \Sigma^{-1} \alpha \, 1' \Sigma^{-1} \} X, \tag{13.3}$$

where

$$\Delta = (\alpha' \Sigma^{-1} \alpha)(1' \Sigma 1) - (\alpha' \Sigma^{-1} 1)^2. \tag{13.4}$$

The variance and the covariance of these estimators are given as

$$\text{Var}(\hat{\mu}) = \frac{\sigma^2 (\alpha' \Sigma^{-1} \alpha)}{\Delta}, \tag{13.5}$$

$$\text{Var}(\hat{\sigma}) = \frac{\sigma^2 (1' \Sigma^{-1} 1)}{\Delta} \tag{13.6}$$

and

$$\text{Cov}(\hat{\mu}, \hat{\sigma}) = -\frac{\sigma^2 (\alpha' \Sigma^{-1} 1)}{\Delta}. \tag{13.7}$$

Note that for any symmetric distribution

$$\alpha_j = -\alpha_{n-j+1}, 1' \Sigma^{-1} \alpha = \alpha' \Sigma^{-1} 1 = 0$$

and

$$\Delta = (\alpha' \Sigma^{-1} \alpha)(1' \Sigma^{-1} 1).$$

Hence the best linear unbiased estimates of μ and σ for the symmetric case are

$$\hat{\mu}^* = \frac{1' \Sigma^{-1} X}{1' \Sigma^{-1} 1}, \tag{13.8}$$

$$\hat{\sigma}^* = \frac{\alpha' \Sigma^{-1} X}{\alpha' \Sigma^{-1} \alpha} \tag{13.9}$$

and the corresponding covariance of the estimators is zero and the their variances are given as

$$\text{Var}(\hat{\mu}^*) = \frac{\sigma^2}{1' \Sigma^{-1} 1} \tag{13.10}$$

and

$$\text{Var}\,(\widehat{\sigma}^*) = \frac{\sigma^2}{\alpha' \Sigma^{-1} \alpha}. \qquad (13.11)$$

We can use the above formulas to obtain the MVLUEs of the location and scale parameters for any distribution numerically provided the variances of the order statistics exist. For some distributions the MVLUEs of the location and scale parameters can be expressed in simplified form.

The following Lemma (see Garybill (1983), p. 198) will be useful to find the inverse of the covariance matrix.

Lemma 13.1. *Let* $\Sigma = (\sigma_{r,s})$ *be* $n \times n$ *matrix with elements, which satisfy the relation*

$$\sigma_{rs} = \sigma_{sr} = c_r d_s, \quad 1 \leqslant r,\ s \leqslant n,$$

for some positive $c_1, \ldots, c_n$ *and* $d_1, \ldots, d_n$. *Then its inverse*

$$\Sigma^{-1} = (\sigma^{r,s})$$

has elements given as follows:

$$\sigma^{1,1} = \frac{c_2}{c_1(c_2 d_1 - c_1 d_2)},$$

$$\sigma^{n,n} = \frac{d_{n-1}}{d_n(c_n d_{n-1} - c_{n-1} d_n)},$$

$$\sigma^{k+1,k} = \sigma^{k,k+1} = -\frac{1}{c_{k+1} d_k - c_k d_{k+1}},$$

$$\sigma^{k,k} = \frac{c_{k+1} d_{k-1} - c_{k-1} d_{k+1}}{(c_k d_{k-1} - c_{k-1} d_k)(c_{k+1} d_k - c_k d_{k+1})}, \quad k = 2, \ldots, n-1,$$

and

$$\sigma^{i,j} = 0, \quad \text{if } |i - j| > 1.$$

Example 13.1. Suppose $X_1, X_2, \ldots, X_n$ are n independent and identically distributed uniform random variables with pdf $F(x)$ given as follows:

$$f(x) = \begin{cases} \dfrac{1}{\sigma}, & \text{if } \mu - \dfrac{\sigma}{2} \leqslant x \leqslant \mu + \dfrac{\sigma}{2}, \\ & \text{where } -\infty < \mu < \infty,\ \sigma > 0, \\ 0, & \text{otherwise.} \end{cases}$$

From results of chapter 8 we have moments of the uniform order statistics:

$$E(X_{r,n}) = \mu + \sigma\left(\frac{r}{n+1} - \frac{1}{2}\right),$$

$$\text{Var}\,(X_{r,n}) = \frac{r(n-r+1)}{(n+1)^2(n+2)}\sigma^2, \quad r = 1, 2, \ldots, n,$$

and

$$\text{Cov}(X_{r,n}, X_{s,n}) = \frac{r(n-s+1)}{(n+1)^2(n+2)}\sigma^2, \quad 1 \leqslant r \leqslant s \leqslant n.$$

We can write

$$\text{Cov}(X_{r,n}, X_{s,n}) = \sigma^2 c_r d_s, \quad 1 \leqslant r \leqslant s \leqslant n,$$

where

$$c_r = \frac{r}{(n+1)^2}, \quad 1 \leqslant r \leqslant n,$$

and

$$d_s = \frac{n-s+1}{n+2}, \quad 1 \leqslant s \leqslant n.$$

Using Lemma 13.1, we obtain that

$$\sigma^{j,j} = 2(n+1)(n+2), \quad j = 1, 2, \ldots, n,$$
$$\sigma^{i,j} = -(n+1)(n+2), \quad j = i+1, \ i = 1, 2, \ldots, n-1,$$
$$\sigma^{i,j} = 0, \quad |i-j| > 1.$$

It can easily be verified that

$$1'\Sigma^{-1} = ((n+1)(n+2), 0, 0, \ldots, 0, (n+1)(n+2)),$$

$$1'\Sigma^{-1}1 = 2(n+1)(n+2),$$

$$1'\Sigma^{-1}\alpha = 0,$$

$$\alpha'\Sigma^{-1} = \left(-\frac{(n+1)(n+2)}{2}, 0, 0, \ldots, 0, \frac{(n+1)(n+2)}{2}\right)$$

and

$$\alpha'\Sigma^{-1}\alpha = \frac{(n-1)(n+2)}{2}.$$

Thus, the MVLUEs of the location and scale parameters μ and σ are

$$\widehat{\mu} = \frac{1'\Sigma^{-1}X}{1'\Sigma^{-1}1} = \frac{X_{1,n} + X_{n,n}}{2}$$

and

$$\widehat{\sigma} = \frac{\alpha'\Sigma^{-1}X}{\alpha'\Sigma^{-1}\alpha} = \frac{(n+1)(X_{n,n} - X_{1,n})}{n-1}.$$

The corresponding covariance of the estimators is zero and their variances are

$$\text{Var}(\widehat{\mu}) = \frac{\sigma^2}{1'\Sigma^{-1}1} = \frac{\sigma^2}{2(n+1)(n+2)}$$

and

$$\text{Var}(\widehat{\sigma}) = \frac{\sigma^2}{\alpha'\Sigma^{-1}\alpha} = \frac{2\sigma^2}{(n-1)(n+2)}.$$

Example 13.2. Suppose that $X_1, X_2, \ldots, X_n$ are n independent and identically distributed exponential random variables with the probability density function, given as

$$f(x) = \begin{cases} \left(\dfrac{1}{\sigma}\right) \exp\left(-\dfrac{x-\mu}{\sigma}\right), & -\infty < \mu < x < \infty, \ 0 < \sigma < \infty, \\ 0, & \text{otherwise.} \end{cases}$$

From chapter 8 we have means, variances and covariances of the exponential order statistics:

$$E(X_{r,n}) = \mu + \sigma \sum_{j=1}^{r} \frac{1}{n-j+1},$$

$$\text{Var}(X_{r,n}) = \sigma^2 \sum_{j=1}^{r} \frac{1}{(n-j+1)^2}, \quad r = 1, 2, \ldots, n,$$

and

$$\text{Cov}(X_{r,n}, X_{s,n}) = \sigma^2 \sum_{j=1}^{r} \frac{1}{(n-j+1)^2}, \quad 1 \leqslant r \leqslant s \leqslant n.$$

One can write that

$$\text{Cov}(X_{r,n}, X_{s,n}) = \sigma^2 c_r d_s, \quad 1 \leqslant r \leqslant s \leqslant n,$$

where

$$c_r = \sum_{j=1}^{r} \frac{1}{(n-j+1)^2}, \quad 1 \leqslant r \leqslant n,$$

and

$$d_s = 1, \quad 1 \leqslant s \leqslant n.$$

Using Lemma 13.1, we obtain (see also example 13.2) that

$$\sigma^{j,j} = (n-j)^2 + (n-j+1)^2, \quad j = 1, 2, \ldots, n,$$

$$\sigma^{j+1,j} = \sigma^{j,j+1} = (n-j)^2, \quad j = 1, 2, \ldots, n-1,$$

and

$$\sigma^{i,j} = 0, \text{ if } |i-j| > 1, \ i, j = 1, 2, \ldots, n.$$

It can easily be shown that

$$1'\Sigma^{-1} = (n^2, 0, 0, \ldots, 0), \alpha'\Sigma^{-1} = (1, 1, \ldots, 1)$$

and

$$\Delta = n^2(n-1).$$

The MVLUEs of the location and scale parameters of μ and σ are respectively

$$\widehat{\mu} = \frac{nX_{1,n} - \overline{X}_n}{n-1}$$

and

$$\widehat{\sigma} = \frac{n(\overline{X}_n - X_{1,n})}{n-1}.$$

The corresponding variances and the covariance of the estimators are

$$\text{Var}\,(\widehat{\mu}) = \frac{\sigma^2}{n(n-1)},$$

$$\text{Var}\,(\widehat{\sigma}) = \frac{\sigma^2}{n-1}$$

and

$$\text{Cov}\,(\widehat{\mu}, \widehat{\sigma}) = -\frac{\sigma^2}{n(n-1)}.$$

Exercise 13.1. Suppose that $X_1, X_2, \ldots, X_n$ are n independent random variables having power function distribution with the pdf $f(x)$ given as

$$f(x) = \begin{cases} \dfrac{\gamma}{\sigma}\left(\dfrac{x-\mu}{\sigma}\right)^{\gamma-1}\gamma, & -\infty < \mu < x < \mu + \sigma, \\ & \text{where } 0 < \sigma < \infty \text{ and } 0 < \gamma < \infty, \\ 0, & \text{otherwise.} \end{cases}$$

Find MVLUEs of the parameters of μ and σ.

Exercise 13.2. Suppose that $X_1, X_2, \ldots, X_n$ are n independent and identically distributed Pareto random variables with pdf $f(x)$, which is given as follows:

$$f(x) = \begin{cases} \dfrac{\gamma}{\sigma}\left(1 + \dfrac{x-\mu}{\sigma}\right)^{-1-\gamma}, & \mu < x < \infty, \\ & \text{here } 0 < \sigma < \infty \text{ and } 0 < \gamma < \infty, \\ 0, & \text{otherwise.} \end{cases}$$

Show that the MVLUEs of parameters μ and σ have the form

$$\widehat{\mu} = X_{1,n} - (c_1 - 1)\widehat{\sigma}$$

and

$$\widehat{\sigma} = M_2\left[\sum_{i=1}^{n-1} -P_i X_{i,n} + \sum_{i=1}^{n-1} P_i X_{n,n}\right],$$

where

$$M_2 = \left(c_n \sum_{i=1}^{n-1} P_i - \sum_{i=1}^{n-1} c_i P_i \right)^{-1},$$

with

$$P_1 = D - (\gamma + 1)d_1,$$
$$P_j = -(\gamma + 1)d_j, \quad j = 2, \dots, n-1,$$
$$P_n = (\gamma - 1)d_n$$

and

$$D = (\gamma + 1) \sum_{i=1}^{n-1} d_i - (\gamma - 1)d_n.$$

The corresponding variances and the covariance of the estimators are

$$\operatorname{Var}(\widehat{\mu}) = E\sigma^2,$$

$$\operatorname{Var}(\widehat{\sigma}) = \left((n\gamma - 1)^2 E - 1 \right) \sigma^2$$

and

$$\operatorname{Cov}(\widehat{\mu}, \widehat{\sigma}) = -\frac{(n\gamma - 1)(n\gamma - 2) - E}{(n\gamma - 2)E} \sigma^2,$$

where

$$E = n\gamma(\gamma - 2) - \frac{(n\gamma - 2)^2}{n\gamma - 2 - D}.$$

Exercise 13.3. Suppose that $X_1, X_2, \dots, X_n$ are n independent and identically distributed from uniform distribution with pdf $f(x)$, which is given as

$$f(x) = \begin{cases} \dfrac{1}{\sigma}, & \mu \leqslant x \leqslant \mu + \sigma, \\ 0, & \text{otherwise.} \end{cases}$$

Show that the MVLUEs of parameters μ and σ have the form

$$\widehat{\mu} = \frac{nX_{1,n} - X_{n,n}}{n-1}$$

and

$$\widehat{\sigma} = \frac{n+1}{n-1}(X_{n,n} - X_{1,n}).$$

Prove also that the corresponding variances and the covariance of the estimators are

$$\operatorname{Var}(\widehat{\mu}) = \frac{n\sigma^2}{(n+2)(n^2 - 1)},$$

$$\operatorname{Var}(\widehat{\sigma}) = \frac{2\sigma^2}{(n+2)(n-1)}$$

and

$$\text{Cov}\,(\widehat{\mu},\widehat{\sigma}) = -\frac{\sigma^2}{(n+2)(n-1)}.$$

Up to now in this chapter we investigated the case of distributions with two unknown parameters. Further, we will study MVLUEs in the situations, when the initial distribution has one unknown parameter only.

Suppose that X has an absolutely continuous distribution function of the form $F(x/\sigma)$, where $\sigma > 0$ is an unknown scale parameter. Further assume that

$$E(X_{r,n}) = \alpha_r\sigma, \quad r = 1, 2, \ldots, n,$$

$$\text{Var}\,(X_{r,n}) = V_{rr}\sigma^2, \quad r = 1, 2, \ldots, n,$$

$$\text{Cov}\,(X_{r,n}, X_{s,n}) = V_{rs}\sigma^2, \quad 1 \leqslant r < s \leqslant n.$$

Let

$$\mathbf{X}' = (X_{1,n}, X_{2,n}, \ldots, X_{n,n}).$$

Then we can write

$$E(\mathbf{X}) = \sigma\boldsymbol{\alpha} \tag{13.12}$$

with

$$\alpha' = (\alpha_1, \alpha_2, \ldots, \alpha_n)$$

and

$$\text{Var}\,(\mathbf{X}) = \sigma^2\boldsymbol{\Sigma},$$

where $\boldsymbol{\Sigma}$ is a matrix with elements $V_{rs}, 1 \leqslant r \leqslant s \leqslant n$.

Then the MVLUE of the scale parameter σ is given as

$$\widehat{\sigma} = \frac{\alpha'\Sigma^{-1}X}{\alpha'\Sigma^{-1}\alpha} \tag{13.13}$$

and

$$\text{Var}\,\widehat{\sigma} = \frac{\sigma^2}{\alpha'\Sigma^{-1}\alpha}. \tag{13.14}$$

Example 13.3. Suppose that $X_1, X_2, \ldots, X_n$ are n independent and identically distributed exponential random variables with pdf given as

$$f(x) = \begin{cases} \left(\dfrac{1}{\sigma}\right)\exp\left(-\dfrac{x}{\sigma}\right), & x > 0, \text{ where } 0 < \sigma < \infty, \\ 0, & \text{otherwise.} \end{cases}$$

Recalling example 13.2, we can write that

$$E(X_{r,n}) = \sigma \sum_{j=1}^{r} \frac{1}{n-j+1},$$

$$\text{Var}(X_{r,n}) = \sigma^2 \sum_{j=1}^{r} \frac{1}{(n-j+1)^2}, \quad r = 1, 2, \dots, n,$$

and

$$\text{Cov}(X_{r,n}, X_{s,n}) = \sigma^2 \sum_{j=1}^{r} \frac{1}{(n-j+1)^2}, \quad 1 \leqslant r \leqslant s \leqslant n.$$

In this situation

$$\text{Cov}(X_{r,n}, X_{s,n}) = \sigma^2 c_r d_s, \quad 1 \leqslant r \leqslant s \leqslant n,$$

where

$$c_r = \sum_{j=1}^{r} \frac{1}{(n-j+1)^2}, \quad 1 \leqslant r \leqslant n,$$

and

$$d_s = 1, \quad 1 \leqslant s \leqslant n.$$

Recalling Lemma 13.1, we obtain that

$$\sigma^{j,j} = (n-j)^2 + (n-j+1)^2, \quad j = 1, 2, \dots, n,$$
$$\sigma^{j+1,j} = \sigma^{j,j+1} = -(n-j)^2, \quad j = 1, \ i, j = 1, 2, \dots, n-1,$$

and

$$\sigma^{i,j} = 0 \ \text{for} \ |i-j| > 1, \ i, j = 1, 2, \dots, n.$$

We have in this case

$$\alpha' = \left(\frac{1}{n}, \frac{1}{n} + \frac{1}{n-1}, \dots, \frac{1}{n} + \dots + \frac{1}{2} + 1 \right),$$

$$\alpha' \Sigma^{-1} = (1, 1, \dots, 1)$$

and

$$\alpha' \Sigma^{-1} \alpha = n.$$

Thus, the MVLUE of σ is given as

$$\widehat{\sigma} = \overline{X}$$

and

$$\text{Var}(\widehat{\sigma}) = \frac{\sigma^2}{n}.$$

Exercise 13.4. Suppose that $X_1, X_2, \ldots, X_n$ are n independent and identically distributed uniform random variables with pdf $f(x)$, which is given as follows:

$$f(x) = \begin{cases} \dfrac{1}{\sigma}, & 0 < x < \sigma, \text{ where } 0 < \sigma < \infty, \text{ and} \\ 0, & \text{otherwise.} \end{cases}$$

Show that the MVLUE of σ in this case is given as

$$\widehat{\sigma} = \frac{(n+1)X_{n,n}}{n}$$

and

$$\mathrm{Var}(\widehat{\sigma}) = \frac{\sigma^2}{n(n+2)}.$$

Check your solutions

Exercise 13.1 (solution). From exercise 8.4, we have

$$E(X_{r,n}) = \mu + \alpha_r \sigma, \quad 1 \leqslant r \leqslant n,$$

and

$$\mathrm{Cov}(X_{r,n}, X_{s,n}) = c_r d_s \sigma^2, \quad 1 \leqslant r \leqslant s \leqslant n,$$

where

$$c_r = \alpha_r = \frac{\Gamma(n+1)\,\Gamma(r+\gamma^{-1})}{\Gamma(n+1+\gamma^{-1})\,\Gamma(r)}, \quad r = 1, 2, \ldots, n,$$

$$d_r = \frac{1}{\alpha_r \beta_r} - \alpha_r,$$

and

$$\beta_r = \frac{\Gamma(n+1+2\gamma^{-1})}{\Gamma(n+1)}\,\frac{\Gamma(r)}{\Gamma(r+2\gamma^{-1})}, \quad r = 1, 2, \ldots, n.$$

Using Lemma 13.1, one can show (see also Kabir and Ahsanullah (1974)) that

$$\sigma^{i,i} = \{(\gamma(i-1)+1)^2 + i\gamma(i\gamma+2)\}\beta_i, \quad i = 1, 2, \ldots, n,$$

$$\sigma^{i+1,i} = \sigma^{i,i+1} = -i(i\gamma+1)\beta_j, \quad i = 1, 2, \ldots, n-1,$$

and

$$\sigma^{i,j} = 0, \quad \text{if } |i-j| > 1.$$

Thus, the MVLUEs of the location and scale parameters μ and σ are respectively

$$\widehat{\mu} = M_1 \left[\sum_{i=1}^{n-1} P_i X_{i,n} - c_n^{-1} \sum_{i=1}^{n-1} c_i P_i X_{n,n} \right]$$

and

$$\widehat{\sigma} = M_2 \left[\sum_{i=1}^{n-1} -P_i X_{i,n} + \sum_{i=1}^{n-1} P_i X_{n,n} \right],$$

where $M_1 = c_n$ and

$$M_2 = \left(c_n \sum_{i=1}^{n-1} P_i - \sum_{i=1}^{n-1} c_i P_i \right)^{-1},$$

with

$$P_1 = (\gamma+1)\beta_1, \quad P_j = (\gamma-1)\beta_j, \quad j = 2,\ldots,n-1.$$

The corresponding variances and the covariance of the estimators are

$$\mathrm{Var}(\widehat{\mu}) = M_1 \sigma^2,$$
$$\mathrm{Var}(\widehat{\sigma}) = \frac{T_n}{\Delta_n} \sigma^2$$

and

$$\mathrm{Cov}(\widehat{\mu},\widehat{\sigma}) = -\frac{1}{\Delta_n d_n} \sigma^2,$$

where

$$T_n = \sum_{i=1}^{n} P_i$$

and

$$\Delta_n = \frac{c_n}{d_n} T_n - \frac{1}{d_n^2}.$$

Exercise 13.2 (hint). One needs to obtain the following preliminary equalities:

$$E(X_{i,n}) = \mu + (c_i - 1)\sigma, \quad \gamma > \frac{1}{n-i+1}$$

and

$$\mathrm{Cov}(X_{i,n} X_{j,n}) = c_j \left(\frac{1}{c_i d_i} - c_i \right) \sigma^2, \quad 1 \leqslant i \leqslant j \leqslant n,$$

if

$$\gamma > \max\left(\frac{2}{n-i+1}, \frac{1}{n-j+1} \right), \quad i \leqslant j, \; i,j = 1,2,\ldots,n,$$

where

$$c_i = \frac{\Gamma(n+1)\Gamma(n+1-i-\gamma^{-1})}{\Gamma(n+1-\gamma^{-1})\Gamma(n+1-i)}$$

and

$$d_i = \frac{\Gamma(n+1-2\gamma^{-1})\Gamma(n+1-i)}{\Gamma(n+1-i-2\gamma^{-1})\Gamma(n+1)}, \quad i = 1, 2, \ldots, n.$$

Then the using of Lemma 13.1 helps to obtain the necessary results (see also Kulldorf and Vannman (1973)).

Exercise 13.3 (hint). Note that the covariance function Σ here is the same as in Example 13.1. Thus, one can use equalities

$$1'\Sigma^{-1} = ((n+1)(n+2), 0, 0, \ldots, 0, (n+1)(n+2)),$$

and

$$1'\Sigma^{-1}1 = 2(n+1)(n+2),$$

obtained there.

Further, in this case

$$\alpha' = \left(\frac{1}{n+1}, \frac{2}{n+1}, 0, \ldots, 0, \ldots, \frac{n}{n+1}\right).$$

Then, we have

$$\alpha'\Sigma^{-1} = (0, 0, 0, \ldots, 0, (n+1)(n+2)),$$

$$\alpha'\Sigma^{-1}\alpha = n(n+2),$$

$$\alpha'\Sigma^{-1}1 = (n+1)(n+2)$$

and

$$\Delta = (n+2)^2(n^2-1).$$

Now the necessary results follow easily from (13.2)-(13.7).

Exercise 13.4 (solution). We have

$$E(X_{r,n}) = \frac{r}{n+1}\sigma,$$

$$\text{Var}(X_{r,n}) = \frac{r(n-r+1)}{(n+1)^2(n+2)}\sigma^2, \quad r = 1, 2, \ldots, n.$$

and

$$\text{Cov}\,(X_{r,n}, X_{s,r}) = \frac{r(n-s+1)}{(n+1)^2(n+2)}\,\sigma^2, \quad 1 \leqslant r \leqslant s \leqslant n.$$

Following example 13.1, we obtain that

$$\alpha' \Sigma^{-1} = (0, 0, \ldots, (n+1)(n+2))$$

and

$$\alpha' \Sigma^{-1} \alpha = n(n+2).$$

Now it follows from (13.13) and (13.14) that

$$\widehat{\sigma} = \frac{(n+1)X_{n,n}}{n}$$

and

$$\text{Var}\,(\widehat{\sigma}) = \frac{\sigma^2}{n(n+2)}\,.$$

Chapter 14

Minimum Variance Linear unbiased Estimators and Predictors Based on Censored Samples

We consider the case, when some smallest and largest observations are missing. In this situation we construct the Minimum Variance Linear unbiased Estimators for location and scale parameters. We discuss also the problem, how to find the best (in the sense of minimum variance) linear unbiased predictor of order statistic $X_{s,n}$, based on given observations $X_{1,n}, X_{2,n}, \ldots, X_{r,n}$, where $1 \leqslant r \leqslant s \leqslant n$.

In the previous chapter we investigated MVLUEs in the case, when all n observations are available for a statistician. Suppose now that the smallest r_1 and largest r_2 of these observations are lost and we can deal with order statistics

$$X_{r_1+1,n} \leqslant \cdots \leqslant X_{n-r_2,n}.$$

We will consider here the minimum variance linear unbiased estimators of the location and scale parameters based on the given elements of the variational series.

Suppose that X has an absolutely continuous distribution function of the form

$$F\left(\frac{x-\mu}{\sigma}\right), \quad -\infty < \mu < \infty, \ \sigma > 0.$$

Further, assume that

$$E(X_{r,n}) = \mu + \alpha_r \sigma,$$
$$\mathrm{Var}(X_{r,n}) = V_{rr}\sigma^2, \quad r_1 + 1 \leqslant r \leqslant n - r_2,$$
$$\mathrm{Cov}(X_{r,n}, X_{s,n}) = V_{rs}\sigma^2, \quad r_1 + 1 \leqslant r, s \leqslant n - r_2.$$

Let

$$\mathbf{X}' = (X_{r_1+1,n}, \ldots, X_{n-r_2,n}).$$

Then we can write that

$$E(\mathbf{X}) = \mu \mathbf{1} + \sigma \boldsymbol{\alpha},$$

M. Ahsanullah et al., *An Introduction to Order Statistics*,
Atlantis Studies in Probability and Statistics 3,
DOI: 10.2991/978-94-91216-83-1_14, © Atlantis Press 2013

with

$$\mathbf{1} = (1, 1, \ldots, 1)',$$

$$\alpha = (\alpha_{r_1+1}, \ldots, \alpha_{n-r_2})',$$

and

$$\mathrm{Var}\,(\mathbf{X}) = \sigma^2 \mathbf{\Sigma},$$

where $\mathbf{\Sigma}$ is a matrix $(n - r_2 - r_1) \times (n - r_2 - r_1)$ having elements V_{rs}, $r_1 < r, s \leqslant n - r_2$.
Then to obtain MVLUEs of the location and scale parameters μ and σ based on the order statistics

$$\mathbf{X}' = (X_{r_1+1,n}, \ldots, X_{n-r_2,n}),$$

a statistician must use the following formulae, which are similar to ones, given in the previous chapter:

$$\widehat{\mu}^* = \frac{1}{\Delta}\{\alpha'\Sigma^{-1}\alpha\mathbf{1}'\Sigma^{-1} - \alpha'\Sigma^{-1}\mathbf{1}\,\alpha'\Sigma^{-1}\}X \tag{14.1}$$

and

$$\widehat{\sigma}^* = \frac{1}{\Delta}\{\mathbf{1}'\Sigma^{-1}\mathbf{1}\,\alpha'\Sigma^{-1} - \mathbf{1}'\Sigma^{-1}\alpha\mathbf{1}'\Sigma^{-1}\}X, \tag{14.2}$$

where

$$\Delta = (\alpha'\Sigma^{-1}\alpha)(\mathbf{1}'\Sigma^{-1}\mathbf{1}) - (\alpha'\Sigma^{-1}\mathbf{1})^2. \tag{14.3}$$

The variances and the covariance of these estimators are given as

$$\mathrm{Var}\,(\widehat{\mu}^*) = \frac{\sigma^2(\alpha'\Sigma^{-1}\alpha)}{\Delta}, \tag{14.4}$$

$$\mathrm{Var}\,(\widehat{\sigma}^*) = \frac{\sigma^2(\mathbf{1}'\Sigma^{-1}\mathbf{1})}{\Delta} \tag{14.5}$$

and

$$\mathrm{Cov}\,(\widehat{\mu}^*, \widehat{\sigma}^*) = -\frac{\sigma^2(\alpha'\Sigma^{-1}\mathbf{1})}{\Delta}. \tag{14.6}$$

Example 14.1. Consider a uniform distribution with cumulative distribution function as

$$F(x) = \frac{2x - 2\mu + \sigma}{2\sigma}, \quad \mu - \frac{\sigma}{2} < x < \mu + \frac{\sigma}{2},$$

where $-\infty < \mu < \infty$ and $\sigma > 0$. Suppose that the smallest r_1 and the largest r_2 observations $(r_1 + r_2 > 1)$ are missing. It can be shown (with some changes in calculations with respect to ones, given in example 13.1) that the inverse Σ^{-1} of the corresponding covariance matrix is

$$\Sigma^{-1} = (n+1)(n+2) \begin{pmatrix} \frac{r_1+2}{r_1+1} & -1 & 0 & 0 & 0 & \cdots & 0 \\ -1 & 2 & -1 & 0 & 0 & \cdots & 0 \\ 0 & -1 & 2 & -1 & 0 & \cdots & 0 \\ 0 & 0 & -1 & 2 & -1 & \cdots & 0 \\ \vdots & \vdots & \vdots & \vdots & \vdots & \ddots & 0 \\ 0 & 0 & 0 & 0 & 0 & \cdots & -1 \\ 0 & 0 & 0 & 0 & 0 & \vdots & \frac{r_2+2}{r_2+1} \end{pmatrix}.$$

Due to (14.1) and (14.2) the MVLUEs of μ and σ are respectively

$$\widehat{\mu}^* = \frac{(n-2r_2-1)X_{r_1+1,n} + (n-2r_1-1)X_{n-r_2,n}}{2(n-r_1-r_2-1)}$$

and

$$\widehat{\sigma}^* = \frac{n+1}{n-r_1-r_2-1}(X_{n-r_2,n} - X_{r_1+1,n}).$$

The variances and the covariance of the estimators are

$$\text{Var}(\widehat{\mu}^*) = \frac{(r_1+1)(n-2r_2-1) + (r_2+1)(n-2r_1-1)}{4(n+2)(n+1)(n-r_1-r_2-1)}\sigma^2,$$

$$\text{Var}(\widehat{\sigma}^*) = \frac{r_1+r_2+2}{(n+2)(n-r_1-r_2-1)}\sigma^2,$$

and

$$\text{Cov}(\widehat{\mu}^*,\widehat{\sigma}^*) = \frac{1}{2(n+1)(n+2)}\Big[(n-2r_1-1)(r_2+1)(n-r_2)$$

$$-(n-2r_2-1)(r_1+1)(n-r_1) - 2(r_2-r_1)(r_1+1)(r_2+1)\Big].$$

Note that if $r_1 = r_2 = r$, then

$$\widehat{\mu}^* = \frac{X_{r+1,n} + X_{n-r,n}}{2}\widehat{\sigma}^*,$$

$$\widehat{\sigma}^* = \frac{n+1}{n-2r-1}(X_{n-r,n} - X_{r+1,n})$$

and

$$\text{cov}(\widehat{\mu}^*,\widehat{\sigma}^*) = 0.$$

Exercise 14.1. Suppose that $X_1, X_2, \ldots, X_n$ are n independent and identically distributed as an exponential distribution with cdf

$$F(x) = 1 - \exp\left(\frac{x-\mu}{\sigma}\right), \quad -\infty < \mu < x < \infty, \ 0 < \sigma < \infty.$$

Assume again that the smallest r_1 and the largest r_2 observations are missing. Show that the MVLUEs of σ and μ, based on order statistics

$$X_{r_1+1,n}, \ldots, X_{n-r_2,n},$$

are

$$\widehat{\sigma}^* = \frac{1}{n-r_2-r_1-1}\left\{ \sum_{i=r_1+1}^{n-r_2} X_{i,n} - (n-r_1)X_{r_1+1,n} + r_2 X_{n-r_2,n} \right\}$$

and

$$\widehat{\mu}^* = X_{r_1+1,n} - \alpha_{r_1+1}\widehat{\sigma}^*,$$

where

$$\alpha_{r_1+1} = \frac{1}{\sigma}E(X_{r_1+1,n} - \mu) = \sum_{i=1}^{r_1+1} \frac{1}{n-i+1}.$$

Prove also that the variances and the covariance of the estimators are

$$\mathrm{Var}\,(\widehat{\mu}^*) = \sigma^2 \left[\frac{\alpha_{r_1+1}^2}{n-r_2-r_1-1} + \sum_{i=1}^{r_1+1} \frac{1}{(n-i+1)^2} \right],$$

$$\mathrm{Var}\,(\widehat{\sigma}^*) = \frac{\sigma^2}{n-r_1-r_2-1}$$

and

$$\mathrm{Cov}\,(\widehat{\mu}^*, \widehat{\sigma}^*) = -\frac{\alpha_r \sigma^2}{n-r_2-r_1-1}.$$

We will discuss the so-called Minimum Variance Linear unbiased Predictors (MVLUPs).

Suppose that $X_{1,n}, X_{2,n}, \ldots, X_{r,n}$ are r ($r < n$) order statistics from a distribution with location and scale parameter μ and σ respectively. Then the best (in the sense of minimum variance) linear predictor $\widehat{X}_{s,n}$ of $X_{s,n}$ ($r < s \leqslant n$) is given as

$$\widehat{X}_{s,n} = \widehat{\mu} + \alpha_s \widehat{\sigma} + W'V^{-1}(X - \widehat{\mu}1 - \widehat{\sigma}\alpha),$$

where $\widehat{\mu}$ and $\widehat{\sigma}$ are MVLUEs of μ and σ respectively, based on

$$\mathbf{X}' = (X_{1,n}, X_{2,n}, \ldots, X_{r,n}),$$

$$\alpha_s = \frac{E(X_{s,n} - \mu)}{\sigma}$$

and

$$W' = (W_1, W_2, \ldots, W_r),$$

where

$$W_i = \mathrm{Cov}(X_{i,n}, X_{s,n}), \quad i = 1, 2, \ldots, r.$$

Here V^{-1} is the inverse of the covariance matrix of $\mathbf{X}$.

Example 14.2. Suppose that

$$X_{1,n}, X_{2,n}, \ldots, X_{r,n} \quad (1 < r < n)$$

are order statistics from a uniform distribution with cdf $F(x)$ given as

$$F(x) = \frac{2x - 2\mu + \sigma}{2\sigma}, \quad \mu - \frac{\sigma}{2} < x < \mu + \frac{\sigma}{2},$$

where

$$\sigma > 0, \quad -\infty < \mu < \infty.$$

Then we have (see examples 13.1 and 14.1)

$$\alpha_i = \frac{2i - 1}{2(n+1)}, \quad i = 1, 2, \ldots, r, s,$$

$$\widehat{\mu} = \frac{(2r - n - 1)X_{1,n} + (n-1)X_{r,n}}{2(r-1)}$$

and

$$\widehat{\sigma} = \frac{n+1}{r-1}(X_{r,n} - X_{1,n}).$$

Further,

$$W_i = \frac{i(n - s + 1)}{(n+1)^2(n+2)},$$

$$V^{-1} = (n+1)(n+2)\begin{pmatrix}
2 & -1 & 0 & 0 & 0 & \cdots & 0 \\
-1 & 2 & -1 & 0 & 0 & \cdots & 0 \\
0 & -1 & 2 & -1 & 0 & \cdots & 0 \\
0 & 0 & -1 & 2 & -1 & \cdots & 0 \\
\vdots & \vdots & \vdots & \vdots & \vdots & \ddots & 0 \\
0 & 0 & 0 & 0 & 0 & \cdots & -1 \\
0 & 0 & 0 & 0 & 0 & \cdots & \frac{n-r+2}{n-r+1}
\end{pmatrix}$$

(compare with matrix Σ^{-1} in example 14.1 under $r_1 = 0$ and $r_2 = n - r$) and thus

$$W'V^{-1} = \left(0, 0, \ldots, \frac{n-s+1}{n-r+1} \right).$$

Hence,

$$\widehat{X}_{s,n} = \widehat{\mu} + \alpha_s \widehat{\sigma} + \frac{n-s+1}{n-r+1}(\widehat{X}_{r,n} - \widehat{\mu} - \alpha_r \widehat{\sigma}).$$

Exercise 14.2. Suppose that

$$X_{1,n}, X_{2,n}, \ldots, X_{r,n} \quad (1 < r < n)$$

are order statistics from an exponential distribution with cdf $F(x)$ given as

$$F(x) = 1 - \exp\left\{ -\frac{x-\mu}{\sigma} \right\}, \quad x > \mu,$$

where

$$-\infty < \mu < \infty, \quad \sigma > 0.$$

Prove that the MVLUP of order statistic $X_{s,n}$ has the form

$$\widehat{X}_{s,n} = \widehat{\mu}^* + \alpha_s \widehat{\sigma}^* + (X_{r,n} - \widehat{\mu}^* - \alpha_r \widehat{\sigma}^*) =$$

$$X_{r,n} + (\alpha_s - \alpha_r)\widehat{\sigma},$$

where

$$\widehat{\mu}^* = X_{1,n} - \alpha_1 \widehat{\sigma}_1,$$

$$\alpha_k = \frac{1}{\sigma} E(X_{k,n} - \mu) = \frac{1}{n} + \cdots + \frac{1}{n-k+1}, \quad k = 1, r, s,$$

and

$$\widehat{\sigma}^* = \frac{1}{r-1}\left\{ \sum_{i=1}^{r} X_{i,n} - nX_{1,n} + (n-r)X_{r,n} \right\}.$$

Check your solutions

Exercise 14.1 (solution). In example 13.2 we discussed already MVLUEs for parameters μ and σ of the exponential distribution with cumulative distribution

$$F(x) = 1 - \exp\left(-\frac{x-\mu}{\sigma} \right), \quad -\infty < \mu < x < \infty, \ 0 < \sigma < \infty,$$

in the case, when all observations were available to a statistician.

We can recall that

$$E(X_{r,n}) = \mu + \sigma \sum_{j=1}^{r} \frac{1}{n-j+1},$$

$$\text{Var}(X_{r,n}) = \sigma^2 \sum_{j=1}^{r} \frac{1}{(n-k+1)^2}, \quad r = 1, 2, \ldots, n,$$

and

$$\text{Cov}(X_{r,n}, X_{s,n}) = \sigma^2 \sum_{j=1}^{r} \frac{1}{(n-j+1)^2}, \quad 1 \leqslant r \leqslant s \leqslant n.$$

To obtain MVLUEs for the case, when $r_1 + r_2$ observations are lost, we need to deal with the covariance matrix Σ of size $(n-r_1-r_2)x(n-r_1-r_2)$, elements of which coincide with

$$\text{Cov}(X_{r,n}, X_{s,n}), \quad r_1 < r, s \leqslant n - r_2.$$

One can write that

$$\text{Cov}(X_{r,n}, X_{s,n}) = \sigma^2 c_r d_s, \quad r_1 + 1 \leqslant r \leqslant s \leqslant n - r_2,$$

where

$$c_r = \sum_{j=1}^{r} \frac{1}{(n-j+1)^2}$$

and

$$d_s = 1.$$

To use formulae (14.1)–(14.6) one needs to find the inverse matrix Σ^{-1}. From lemma 13.1, we obtain that

$$\Sigma^{-1} = \begin{pmatrix} (n-r_1-1)^2 + \frac{1}{c_{r+1}} & -(n-r_1-1)^2 & \cdots & 0 \\ -(n-r_1-1)^2 & (n-r_1-2)^2 + (n-r_1-1)^2 & \cdots & 0 \\ 0 & -(n-r_1-2)^2 & \cdots & 0 \\ 0 & 0 & \cdots & 0 \\ \vdots & \vdots & \vdots & 0 \\ 0 & 0 & \cdots & -(r_2+1)^2 \\ 0 & 0 & \cdots & (r_2+1)^2 \end{pmatrix},$$

where

$$\sigma^{1,1} = (n-r_1-1)^2 + \frac{1}{c_{r+1}}$$

$$\sigma^{n-r_1-r_2,n-r_1-r_2} = (r_2+1)^2,$$

$$\sigma^{j,j} = (n-r_1-j)^2 + (n-r_1-j+1)^2, \quad j = 2, 3, \ldots, n-r_1-r_2-1,$$

$$\sigma^{j+1,j} = \sigma^{j,j+1} = -(n-r_1-j)^2, \quad j = 1, 2, \ldots, n-r_1-r_2-1,$$

and

$$\sigma^{i,j} = 0 \text{ for } |i-j| > 1, \ i,j = 1,2,\ldots,n-r_1-r_2.$$

Note also that in (14.1)–(14.6) we must use

$$\alpha = (\alpha_{r_1+1},\ldots,\alpha_{n-r_2})',$$

where

$$\alpha_r = \frac{E(X_{r,n}) - \mu}{\sigma} = \sum_{j=1}^{r} \frac{1}{n-j+1}.$$

The simple calculations show that

$$\alpha' \Sigma^{-1} = \left(\frac{\alpha_{r_1+1}}{c_{r_1+1}} - (n-r_1-1), 1, \ldots, 1, r_2+1 \right);$$

$$\alpha' \Sigma^{-1} \alpha = \frac{\alpha_{r_1+1}^2}{c_{r_1+1}} + (n-r_1-r_2-1);$$

$$\alpha' \Sigma^{-1} 1 = \frac{\alpha_{r_1+1}}{c_{r_1+1}};$$

$$1' \Sigma^{-1} 1 = \frac{1}{c_{r_1+1}};$$

$$1' \Sigma^{-1} \alpha = \frac{\alpha_{r_1+1}}{c_{r_1+1}};$$

$$1' \Sigma^{-1} 1 \alpha' \Sigma^{-1} = \frac{1}{A_{r_1+1}} \left(\frac{\alpha_{r_1+1}}{c_{r_1+1}} - (n-r_1-1), 1, \ldots, 1, r_2+1 \right);$$

$$1' \Sigma^{-1} \alpha 1' \Sigma^{-1} = \frac{1}{A_{r_1+1}} \left(\frac{\alpha_{r_1+1}}{c_{r_1+1}}, 0, \ldots, 0 \right);$$

$$\Delta = (\alpha' \Sigma^{-1} \alpha)(1' \Sigma^{-1} 1) - (\alpha \Sigma^{-1} 1)^2 = \frac{n-r_1-r_2-1}{c_{r_1+1}}.$$

Now, we obtain from (14.2) that

$$\widehat{\sigma}^* = \frac{1}{\Delta} \left\{ 1' \Sigma^{-1} 1 \, \alpha' \Sigma^{-1} - 1' \Sigma^{-1} \alpha 1' \Sigma^{-1} \right\} X =$$

$$\frac{1}{n-r_2-r_1-1} \left\{ \sum_{i=r_1+1}^{n-r_2} X_{i,n} - (n-r_1)X_{r_1+1,n} + r_2 X_{n-r_2,n} \right\}.$$

Analogously from (14.1), (14.4)–(14.6) we have the necessary expressions for

$$\widehat{\mu}^*, \text{Var}\,(\widehat{\mu}^*), \text{Var}\,(\widehat{\sigma}^*)$$

and

$$\text{Cov}\,(\widehat{\mu}^*, \widehat{\sigma}^*).$$

Exercise 14.2 (hint). Compare with exercise 14.1 and find that

$$
V^{-1} = \begin{pmatrix}
(n-1)^2+n^2 & -(n-1)^2 & \cdots & 0 \\
-(n-1)^2 & (n-2)^2+(n-1)^2 & \cdots & 0 \\
0 & -(n-2)^2 & \cdots & 0 \\
0 & 0 & \cdots & 0 \\
\vdots & \vdots & \ddots & 0 \\
0 & 0 & \cdots & -(n-r+1)^2 \\
0 & 0 & \cdots & (n-r+1)^2
\end{pmatrix}
$$

Other characteristics, which are necessary to obtain MVLUP can be taken from example 13.2 and exercise 14.1.

Chapter 15

Estimation of Parameters Based on Fixed Number of Sample Quantiles

We continue to investigate best (in the sense of minimum variance) unbiased estimators of the location and scale parameters. In this chapter MVLUEs, based on some fixed order statistics, are discussed

First of all we recall the definitions of a quantile of order λ and a sample quantile of order λ, $0 < \lambda < 1$, given in chapter 3.

A value x_λ is called a quantile of order λ, $0 < \lambda < 1$, if

$$P(X < x_\lambda) < \lambda \leqslant P(X \leqslant x_{\lambda'}).$$

If X has a continuous cdf F, then any solution x_λ of the equation $F(x_\lambda) = \lambda$, is a quantile of order λ.

Following definitions given in chapter 3 we determine the sample quantile of order λ, $0 < \lambda < 1$, as $X_{[n\lambda]+1,n}$. For the sake of simplicity instead of $X_{[n\lambda]+1,n}$ we will use notation x_λ^s.

Consider a set of real numbers

$$0 < \lambda_1 < \lambda_2 < \cdots < \lambda_k < 1,$$

and let

$$x_{\lambda_1}^s, x_{\lambda_2}^s, \ldots, x_{\lambda_k}^s$$

be the corresponding sample quantiles. Mosteller (1946) showed that the joint distribution of normalized sample quantiles

$$\sqrt{n}(x_{\lambda_1}^s - x_{\lambda_1}), \sqrt{n}(x_{\lambda_2}^s - x_{\lambda_2}), \ldots, \sqrt{n}(x_{\lambda_k}^s - x_{\lambda_k}),$$

M. Ahsanullah et al., *An Introduction to Order Statistics*,
Atlantis Studies in Probability and Statistics 3,
DOI: 10.2991/978-94-91216-83-1_15, © Atlantis Press 2013

corresponding to a distribution with pdf $f(x)$, tends, as $n \to \infty$, to a k-dimensional normal distribution with zero means and covariance matrix

$$
\begin{pmatrix}
\dfrac{\lambda_1(1-\lambda_1)}{(f(x_{\lambda_1}))^2} & \dfrac{\lambda_1(1-\lambda_2)}{f(x_{\lambda_1})f(x_{\lambda_2})} & \cdots & \dfrac{\lambda_1(1-\lambda_k)}{f(x_{\lambda_1})f(x_{\lambda_k})} \\[2ex]
\dfrac{\lambda_1(1-\lambda_2)}{f(x_{\lambda_1})f(x_{\lambda_2})} & \dfrac{\lambda_2(1-\lambda_2)}{(f(x_{\lambda_2}))^2} & \cdots & \dfrac{\lambda_2(1-\lambda_k)}{f(x_{\lambda_2})f(x_{\lambda_k})} \\[2ex]
\vdots & \vdots & \ddots & \vdots \\[2ex]
\dfrac{\lambda_1(1-\lambda_k)}{f(x_{\lambda_1})f(x_{\lambda_k})} & \cdots & \cdots & \dfrac{\lambda_k(1-\lambda_k)}{(f(x_{\lambda_k}))^2}
\end{pmatrix}
$$

In two previous chapters we discussed the best (in the sense of minimum variances) linear unbiased estimators based on the whole set or some censored set of order statistics. When the size n of a sample is large, a statistician can have some technical problems to construct the appropriate estimators based on large number of order statistics. Ogawa (1951) suggested to simplify the problem considering the MVLUEs of the location and scale parameters, which use only fixed number k, say $k = 2$ or $k = 3$, of order statistics. This simplification is of great interest, if n is sufficiently large, and then it is better to solve the problem in terms of k sample quantiles.

The statement of the problem is the following. We have observations, corresponding to a distribution function $F((x - \mu)/\sigma)$. Here $F(x)$ is known and has the probability density function $f(x)$, μ and σ are the location and scale parameters. One or both of the parameters are unknown. We want to construct the MVLUE of the unknown parameter (parameters) based on fixed set

$$(x^s_{\lambda_1}, x^s_{\lambda_2}, \ldots, x^s_{\lambda_k})$$

of sample quantiles, where

$$0 < \lambda_1 < \lambda_2 < \cdots < \lambda_k < 1.$$

To simplify our calculations we suppose that $n \to \infty$ and instead of the joint distribution of sample quantiles, one can use its asymptotic expression, given above.

Ogawa considered some important situations.

Case 1. *MVLUE of the location parameter μ, when the scale parameter σ is known...*

Let $\widehat{\mu}_q$ denote the MVLUE of μ, based on linear combinations of k fixed sample quantiles

$$(x^s_{\lambda_1}, x^s_{\lambda_2}, \ldots, x^s_{\lambda_k}).$$

It turns out that $\widehat{\mu}_q$ has the form

$$\widehat{\mu}_q = \frac{T}{K_1} - \frac{K_3}{K_1}\sigma, \tag{15.1}$$

and the variance of $\widehat{\mu}_q$ behaves as

$$\mathrm{Var}\,(\widehat{\mu}_q) \sim \frac{\sigma^2}{nK_1}, \quad n \to \infty,$$

where

$$T = T(x^s_{\lambda_1}, \ldots, x^s_{\lambda_k}) = \sum_{i=1}^{k+1} \frac{(f(x_{\lambda_i}) - f(x_{\lambda_{i-1}}))\,(f(x_{\lambda_i})x^s_{\lambda_i} - f(x_{\lambda_{i-1}})x^s_{\lambda_{i-1}})}{\lambda_i - \lambda_{i-1}},$$

$$\lambda_0 = 0, \quad \lambda_{k+1} = 1, \quad f(x_{\lambda_0}) = 0, \quad f(x_{\lambda_{k+1}}) = 0,$$

$$K_1 = \sum_{i=1}^{k+1} \frac{(f(x_{\lambda_i}) - f(x_{\lambda_{i-1}}))^2}{\lambda_i - \lambda_{i-1}} \tag{15.2}$$

and

$$K_3 = \sum_{i=1}^{k+1} \frac{(f(x_{\lambda_i}) - f(x_{\lambda_{i-1}}))(f(x_{\lambda_i})x_{\lambda_i} - f(x_{\lambda_{i-1}})x_{\lambda_{i-1}})}{\lambda_i - \lambda_{i-1}}. \tag{15.3}$$

Case 2. *MVLUE of the scale parameter σ, when the location parameter μ is known.*

Let $\widehat{\sigma}_q$ be the MVLUE of σ. It turns out that

$$\widehat{\sigma}_q = \frac{S}{K_2} - \frac{K_3}{K_2}\sigma, \tag{15.4}$$

and

$$\mathrm{Var}\,(\widehat{\sigma}_q) \sim \frac{\sigma^2}{nK_2}, \quad n \to \infty,$$

where

$$S = S(x^s_{\lambda_1}, \ldots, x^s_{\lambda_k})$$
$$= \sum_{i=1}^{k+1} \frac{(f(x_{\lambda_i})x_{\lambda_i} - f(x_{\lambda_{i-1}})x_{\lambda_{i-1}})(f(x_{\lambda_i})x^s_{\lambda_i} - f(x_{\lambda_{i-1}})x^s_{\lambda_{i-1}})}{\lambda_i - \lambda_{i-1}},$$
$$K_2 = \sum_{i=1}^{k+1} \frac{(f(x_{\lambda_i})x_{\lambda_i} - f(x_{\lambda_{i-1}})x_{\lambda_{i-1}})^2}{\lambda_i - \lambda_{i-1}}$$

and K_3 is given in (15.3).

Case 3. *MVLUEs of location and scale parameters, when both μ and σ are unknown.*

Let $\widehat{\mu}_{q0}$ and $\widehat{\sigma}_{q0}$ be the MVLUEs of μ and σ, based on linear combinations of k fixed sample quantiles

$$(x^s_{\lambda_1}, x^s_{\lambda_2}, \ldots, x^s_{\lambda_k})$$

respectively. Then

$$\widehat{\mu}_{q0} = \frac{K_2 T - K_3 S}{\Delta}$$

and

$$\widehat{\sigma}_{q0} = \frac{-K_3 T + K_1 S}{\Delta},$$

where

$$\Delta = K_1 K_2 - K_3^2$$

and the expressions for

$$T = T(x_{\lambda_1}^s, \ldots, x_{\lambda_k}^s),$$
$$S = S(x_{\lambda_1}^s, \ldots, x_{\lambda_k}^s),$$

K_1, K_2 and K_3 are given above.

The variances and the covariance of the estimates $\widehat{\mu}_{q0}$ and $\widehat{\sigma}_{q0}$ behave as follows:

$$\text{Var}(\widehat{\mu}_{q0}) \sim \frac{K_2}{n\Delta} \sigma^2,$$
$$\text{Var}(\widehat{\sigma}_{q0}) \sim \frac{K_1}{n\Delta} \sigma^2$$

and

$$\text{Cov}(\widehat{\mu}_{q0}, \widehat{\sigma}_{q0}) \sim -\frac{K_3}{n\Delta} \sigma^2,$$

as $n \to \infty$.

If the pdf $F(x)$ is symmetric about zero and the location of

$$0 < \lambda_1 < \lambda_2 < \cdots < \lambda_k < 1$$

is symmetric with respect to $1/2$, i.e.

$$\lambda_j + \lambda_{k-j+1} = 1, \quad j = 1, 2, \ldots, k,$$

then $K_3 = 0$ and $\text{Cov}(\widehat{\mu}_{q0}, \widehat{\sigma}_{q0}) = 0$.

Moreover, $\widehat{\mu}_{q0}$ and $\widehat{\sigma}_{q0}$ in this case coincide with $\widehat{\mu}_q$ from case 1 and $\widehat{\sigma}_q$ from case 2 respectively.

Example 15.1. Consider the exponential distribution with pdf

$$f\left(\frac{x}{\sigma}\right), \quad \sigma > 0,$$

where

$$f(x) = \exp(-x), \quad x > 0.$$

Using formulae, given for the case 2, we can show that

$$\widehat{\sigma}_q = \sum_{i=1}^{k} a_j x_{\lambda_j}^s,$$

where

$$a_j = e^{-u_i}\left(\frac{e^{-u_i}u_i - e^{-u_{i-1}}u_{i-1}}{e^{-u_i} - e^{-u_{i-1}}} - \frac{e^{-u_{i+1}}u_i - e^{-u_{i-1}}u_i}{e^{-u_{i+1}} - e^{-u_i}}\right)$$

and

$$U_j = -\log(1 - \lambda_j), \quad j = 1, 2, \ldots, k.$$

The corresponding variance is given as follows:

$$\mathrm{Var}(\widehat{\sigma}_q) \sim \frac{\sigma^2}{nK_2}, \quad n \to \infty,$$

where

$$K_2 = \sum_{i=1}^{k+1} \frac{\left((1 - \lambda_i)\log(1 - \lambda_i) - (1 - \lambda_{i-1})\log(1 - \lambda_{i-1})\right)^2}{\lambda_i - \lambda_{i-1}}. \tag{15.5}$$

For any fixed k it is interesting to find the set $(\lambda_1, \lambda_2, \ldots, \lambda_k)$, which minimizes the asymptotic expression of $\mathrm{Var}(\widehat{\sigma}_q)$ or equivalently gives the maximum value of K_2.

Let $k = 1$. It means that we want to use one sample quantile only to estimate σ. What value of λ_1 must be chosen to obtain the estimate with minimum variance?

If $k = 1$, then

$$K_2 = \frac{1 - \lambda_1}{\lambda_1}\left(-\ln(1 - \lambda_1)\right)^2.$$

It can be shown that K_2 is maximum, when

$$\lambda_1 = 0.7968.$$

In this case

$$K_2 = 0.6476.$$

Exercise 15.1. Consider $k = 2$ in example 15.1 and find the MVLUE of the scale parameter σ, based on two sample quantiles.

Exercise 15.2. For $k = 3$ in example 15.1, show that K_2 takes on its maximum value, when

$$\lambda_1 = 0.5296,$$

$$\lambda_2 = 0.8300$$

and

$$\lambda_3 = 0.9655.$$

Check your solutions

Exercise 15.1 (solution). For $k = 2$ from (15.5) we obtain that

$$K_2 = \sum_{i=1}^{3} \frac{\left((1 - \lambda_i)\log(1 - \lambda_i) - (1 - \lambda_{i-1})\log(1 - \lambda_{i-1})\right)^2}{\lambda_i - \lambda_{i-1}},$$

where $\lambda_0 = 0$ and $\lambda_3 = 1$. The standard procedure enables us to find that K_2 is maximum, when

$$\lambda_1 = 0.6386$$

and

$$\lambda_2 = 0.9266.$$

The maximal value of K_2 is 0.8203.

Exercise 15.2 (hint). For $k = 3$,

$$K_2 = K_2(\lambda_1, \lambda_2, \lambda_3)$$
$$= \sum_{i=1}^{4} \frac{\left((1 - \lambda_i)\log(1 - \lambda_i) - (1 - \lambda_{i-1})\log(1 - \lambda_{i-1})\right)^2}{\lambda_i - \lambda_{i-1}},$$

where $\lambda_0 = 0$ and $\lambda_4 = 1$. Use the standard methods to find the maximum value of K_2. It turns out that $K_2(\lambda_1, \lambda_2, \lambda_3)$ has its maximum, when

$$\lambda_1 = 0.5296,$$
$$\lambda_2 = 0.8300$$
$$\lambda_3 = 0.9655,$$

and

$$K_2(0.5296, 0.8300, 0.9655) = 0.8910.$$

Chapter 16

Order Statistics from Extended Samples

Sometimes besides order statistics $X_{1,n}, \ldots, X_{n,n}$ a statistician may additionally use some order statistics, which are chosen from a variational series $X_{1,n+m} \leqslant \cdots \leqslant X_{n+m,n+m}$. We investigate independence properties of differences $X_{k,n} - X_{l,n+m}$ in the case, when the initial distribution is exponential.

Let $X_1, X_2, \ldots$ be a sequence of independent random variables (r.v.'s) with a common continuous distribution function (d.f.) F and let $X_{1,n} \leqslant \cdots \leqslant X_{n,n}$ be order statistics based on r.v.'s $X_1, \ldots, X_n$, $n = 1, 2, \ldots$. Sometimes one needs to consider a set of order statistics $X_{i(1),n}, \ldots, X_{i(k),n}$ together with some additional r.v.'s $X_{n+1}, \ldots, X_{n+m}$. There are some natural situations, when a statistician has to make inference about unknown parameters or to test some statistical hypotheses based on order statistics taken from a variational series

$$X_{1,n} \leqslant \cdots \leqslant X_{n,n}$$

and an extended variational series

$$X_{1,n+m} \leqslant \cdots \leqslant X_{n+m,n+m}.$$

Imagine that new observations regularly come from a parent distribution but in any time only k maximal values from a recent sample are available for the statistician. In this case he/she needs to know joint distributions of order statistics $X_{m,n}$ and $X_{l,n+m}$ that to test a suggestion that new observations $X_{n+1}, \ldots, X_{n+m}$ arrived from the same distribution as the first n observations did. Very close to this topic are also results for overlapping samples and moving order statistics. Moving samples of size n are used very often in a quality control. Different characteristics based on order statistics for moving samples such as moving ranges or moving midranges among other moving statistics can indicate location and dispersion changes in a time series. Extended and moving samples are the partial cases of overlapping samples. Different applications of order statistics for overlapping samples

M. Ahsanullah et al., *An Introduction to Order Statistics*,
Atlantis Studies in Probability and Statistics 3,
DOI: 10.2991/978-94-91216-83-1_16, © Atlantis Press 2013

were suggested in Blom (1980), Cleveland and Kleiner (1975). Overlapping order statistics arise, for example, in some subclasses of U-statistics. distributions and some moment characteristics for the corresponding order statistics were discussed, for example, in Inagaki (1980) and David and Rogers (1983). Ahsanullah and Nevzorov (1997) investigated independence properties of extended order statistics. They obtained conditional distributions of $P\{X_{k,n} \leqslant x \mid X_{l,n+m} = y\}$ and $P\{X_{l,n+m} \leqslant x \mid X_{k,n} = y\}$ and got some results for the exponential case. For instance, the following examples were given.

Example 16.1. If
$$F(x) = \max\left\{0, 1 - e^{-(x-a)/b}\right\},$$
then for any $1 \leqslant k \leqslant n$, $1 \leqslant l \leqslant n+m$, $n \geqslant 1$, $m \geqslant 1$, random variables
$$T_k = \max(0, X_{k,n} - X_{l,n+m})$$
and $X_{l,n+m}$ are independent.

Remark 16.1. The statement, presented in example 16.1, was proved in chapter 4 for the partial case $m = 0$. It was shown there that if
$$F(x) = \max\left\{0, 1 - e^{-(x-a)/b}\right\},$$
then for any $1 \leqslant k < l \leqslant n$, the difference $X_{l,n} - X_{k,n}$ and order statistic $X_{k,n}$ are independent.

Example 16.2. Let $X_1, X_2, \ldots$ be independent random variables with the same distribution function F and density function f. Then random variables
$$T_k = \max(0, X_{k,n} - X_{l,n+m})$$
and $X_{l,n+m}$ are independent for $l < k + m$ iff
$$F(x) = 1 - \exp\left(-\frac{x-a}{b}\right), \quad x > a,$$
where $b > 0$ and $-\infty < a < \infty$.

In fact, a more general result, than the statement given in example 16.1, is valid for exponential distributions.

Example 16.3. Let
$$F(x) = 1 - \exp\left(-\frac{x-a}{b}\right), \quad x > a.$$
Then a vector of spacings
$$(T_1, \ldots, T_n)$$
does not depend on order statistics
$$X_{1,n+m}, X_{2,n+m}, \ldots, X_{n+m,n+m}.$$
The result presented in example 16.3 implies the following evident corollaries.

Corollary 16.1. *If $m = 0$ then we obtain the well-known independence property of the exponential spacings (see Remark 4.5).*

Corollary 16.2. *Differences*

$$T_2 - T_1, \ldots, T_n - T_{n-1}$$

do not depend on order statistics $X_{1,n+m}, X_{2,n+m}, \ldots, X_{n+m,n+m}$.

Corollary 16.3. *Spacings*

$$X_{n,n} - X_{n-1,n}, X_{n-1,n} - X_{n-2,n}, \ldots, X_{l+2,n} - X_{l+1,n}, X_{l+1,n} - X_{l,n}$$

do not depend on order statistics

$$X_{1,n+m}, X_{2,n+m}, \ldots, X_{l,n+m}, \quad l = 1, 2, \ldots, n-1.$$

Independence properties of extended exponential spacings enable us to obtain joint moments of order statistics $X_{k,n}$ and $X_{l,n+m}$, if $k \geqslant l$.

Exercise 16.1. Let us consider the standard exponential distribution with

$$F(x) = 1 - \exp(-x), \quad x > 0.$$

For $k \geqslant l$ find moments

$$E(X_{k,n} X_{l,n+m})$$

and covariances

$$\mathrm{Cov}\,(X_{k,n}, X_{l,n+m}).$$

The independence of $X_{1,n+m}$ and $X_{1,n} - X_{1,n+m}$ for exponential order statistics enables us to write the expression for the regression

$$\varphi(x) = E(X_{1,n} - X_{1,n+m} \mid X_{1,n+m} = x).$$

It is evident that in this case

$$\varphi(x) = c, \quad x > 0,$$

where c is a positive constant. It turns out (see, for example, equality (1.4.20) in Ahsanullah and Nevzorov (2001)) that in the general situation

$$\varphi(x) = E(X_{1,n} - X_{1,n+m} \mid X_{1,n+m} = x)$$
$$= \frac{nm}{n+m} \frac{1}{(1-F(x))^n} \int_x^\infty v(1-F(v))^{n-1} dF(v) - \frac{mx}{n+m}. \tag{16.1}$$

Due to (16.1) we can prove that the equality

$$\varphi(x) = c, \quad x > d > -\infty,$$

is valid for distribution with exponential tails only.

Exercise 16.2. Prove that $F(x)$ has the form

$$F(x) = 1 - \exp\left(-\frac{x-d}{b}\right), \quad x > d,$$

where $-\infty < d < \infty$ and $b > 0$, if

$$\varphi(x) = c, \quad x > d > -\infty,$$

where c is a positive constant.

Check your solutions

Exercise 16.1 (solution). From corollary 16.3 we know that spacings

$$X_{l,n} - X_{l,n+m}, \ldots, X_{n,n} - X_{l,n+m}$$

do not depend on order statistics

$$X_{1,n+m}, \ldots, X_{l,n+m}.$$

Hence if $r \geqslant l$ then r.v.'s $T_r = (X_{r,n} - X_{l,n+m})$ and $X_{l,n+m}$ are independent. From the equality

$$0 = \mathrm{Cov}\,(T_r, X_{l,n+m}) = \mathrm{Cov}\,(X_{r,n}, X_{l,n+m}) - \mathrm{Var}\,(X_{l,n+m}),$$

one gets that

$$\mathrm{Cov}\,(X_{r,n}, X_{l,n+m}) = \mathrm{Var}\,(X_{l,n+m}).$$

From formulae (8.25) and (8.26) for expectations and variances of exponential order statistics we obtain that

$$EX_{l,n+m} = \frac{1}{n+m} + \frac{1}{n+m-1} + \cdots + \frac{1}{n+m-l+1},$$

$$\mathrm{Var}\,X_{l,n+m} = EX_{l,n+m} - \sum_{k=n+m-l+1}^{n+m} \frac{1}{k^2}.$$

Thus,

$$\mathrm{Cov}\,(X_{r,n}, X_{l,n+m}) = \sum_{k=n+m-l+1}^{n+m} \frac{1}{k} - \sum_{k=n+m-l+1}^{n+m} \frac{1}{k^2}$$

and

$$EX_{r,n}X_{l,n+m} = \mathrm{Cov}\,(X_{r,n}, X_{l,n+m}) + EX_{l,n+m}EX_{r,n}$$

$$= \mathrm{Var}\,X_{l,n+m} + EX_{l,n+m}EX_{r,n}$$

$$= \sum_{k=n+m-l+1}^{n+m} \frac{1}{k} - \sum_{k=n+m-l+1}^{n+m} \frac{1}{k^2} + \sum_{k=n+m-l+1}^{n+m} \frac{1}{k} \sum_{k=n-r+1}^{n} \frac{1}{k}.$$

Exercise 16.2 (solution). Let $R(v) = 1 - F(v)$. Then one can rewrite (16.1) as

$$\varphi(x) = -\frac{mx}{n+m} - \frac{nm}{n+m} \frac{1}{R^n(x)} \int_x^\infty v R^{n-1}(v) dR(v)$$

$$= \frac{m}{n+m} \int_x^\infty \left(\frac{R(u)}{R(x)} \right)^n du. \tag{16.2}$$

We must find a function $R(x)$ such that

$$\varphi(x) = c, \quad x > d.$$

Due to (16.2) we need to solve the equation

$$c(R(x))^n = \frac{m}{n+m} \int_x^\infty (R(u))^n du. \tag{16.3}$$

Let us denote

$$I(x) = \int_x^\infty (R(u))^n du.$$

Then (16.3) can be rewritten as

$$I'(x) = -\frac{m}{c(n+m)} I(x). \tag{16.4}$$

This differential equation has solutions of the form

$$I(x) = \exp\left\{ -\frac{x-d}{b} \right\}, \quad x > d,$$

b being the positive constant. Hence, all possible expressions for $(R(u))$ have the same form. Recalling that $F(x) = 1 - R(x)$ is a distribution function, we obtain that all possible distribution functions $F(x)$, for which the relation $\varphi(x) = c$ is valid for $x > d$, are determined for $x > d \succeq a$:

$$F(x) = 1 - \exp\left(-\frac{x-d}{b} \right), \quad x > d.$$

Chapter 17

Order Statistics and Record Values

The theory of records is tied very closely with the theory of extreme order statistics. The definitions and different properties of record indicators, record times and record values are given here.

Very close to order statistics are the so-called record times and record values. Beginning from the Chandler's (1952) pioneer paper records became very popular in the probability theory and statistics. A lot of monographs (see, for example, Ahsanullah (1995), Arnold, Balakrishnan and Nagaraja (1998) and Nevzorov (2001, 2000)) are devoted to investigation of records.

Let $X_1, X_2, \ldots$ be a sequence of random variables and $X_{1,n} \leqslant \cdots \leqslant X_{n,n}$, $n = 1, 2, \ldots$, be the corresponding order statistics. Denote, for simplicity,

$$M_n = X_{n,n} = \max \{X_1, X_2, \ldots, X_n\}, \quad n = 1, 2, \ldots,$$

and consider the increasing sequence of these sequential maximal values

$$M_1 \leqslant M_2 \leqslant \cdots \leqslant M_n \leqslant M_{n+1} \leqslant \cdots.$$

Let us fix the times, when signs of the strong inequality appear in this sequence. Such times correspond to the situation, when $M_n > M_{n-1}$. It means that $X_n > M_{n-1}$. The random variable X_n, which is more than all previous X's, is called the upper record value. The corresponding index n is named as a record time. Usually X_1 is taken as the first record value. The theory of records is closely connected with the theory of order statistics.

Let us discuss the main problems related to the theory of record values.

Denote $X(1) < X(2) < \cdots$ the record values in the sequence $X_1, X_2, \ldots$ and let $1 = L(1) < L(2) < \cdots$ be the corresponding record times. Introduce also record indicators ξ_n, $n = 1, 2, \ldots$, which take values 0 and 1 and mark the appearance of record values, that is

$$\xi_n = \begin{cases} 1, & \text{if } X_n > M_{n-1}, \\ 0, & \text{otherwise.} \end{cases}$$

M. Ahsanullah et al., *An Introduction to Order Statistics*,
Atlantis Studies in Probability and Statistics 3,
DOI: 10.2991/978-94-91216-83-1_17, © Atlantis Press 2013

As we agreed above, $L(1) = 1$, $X(1) = X_1$ and $\xi_1 = 1$. Note that $\xi_{L(n)} = 1$, $n = 1, 2, \ldots$.

We restrict our consideration by the classical situation, when $X_1, X_2, \ldots$ are independent random variables with a common continuous distribution function F. The fact that X's have the continuous distribution allows us not to take into account the events, when values of some X's coinside. Indeed, the corresponding theory of records is also developed for X's with discrete distributions.

Record indicators

Firstly we discuss the most important properties of the record indicators.

Theorem 17.1. *Let* $X_1, X_2, \ldots$ *be a sequence of independent random variables with a common distribution function* F. *Then record indicators* $\xi_1, \xi_2, \ldots$ *are independent and*

$$P\{\xi_n = 1\} = 1 - P\{\xi_n = 0\} = \frac{1}{n}, \quad n = 1, 2, \ldots. \tag{17.1}$$

Proof. Really,

$$P\{\xi_n = 1\} = P\{X_n > M_{n-1}\} = P\{X_n = M_n\} = P\{X_n = X_{n,n}\}.$$

Since $X_1, X_2, \ldots, X_n$ have a continuous distribution, $P\{X_k = X_m\} = 0$, if $k \neq m$, and hence, due to the symmetry of events $\{X_k = X_{n,n}\}$, $k = 1, 2, \ldots, n$, we get that

$$1 = \sum_{k=1}^{n} P\{X_k = X_{n,n}\} = n P\{X_n = X_{n,n}\} = n P\{\xi_n = 1\}. \tag{17.2}$$

It follows immediately from (17.2) that $P\{\xi_n = 1\} = 1/n$.

All indicators have two values only. Hence to prove that ξ's are independent it is enough to check that

$$P\{\xi_1 = 1, \xi_2 = 1, \ldots, \xi_n = 1\} = \frac{1}{n!}, \quad n = 2, 3, \ldots. \tag{17.3}$$

Indeed,

$$P\{\xi_1 = 1, \xi_2 = 1, \ldots, \xi_n = 1\} = P\{X_1 < X_2 < \cdots < X_n\}$$

and

$$1 = \sum P\{X_{\alpha(1)} < X_{\alpha(2)} < \cdots < X_{\alpha(n)}\}, \tag{17.4}$$

where sum on the RHS of (17.4) is taken over all $n!$ permutations $(\alpha(1), \alpha(2), \ldots, \alpha(n))$ of numbers $1, 2, \ldots, n$. All $n!$ events $\{X_{\alpha(1)} < X_{\alpha(2)} < \cdots < X_{\alpha(n)}\}$ have the same probabilities and each of them (including $P\{X_1 < X_2 < \cdots < X_n\}$) is equal to $1/n!$.

Thus the statement of Theorem 17.1 is proved. $\square$

Exercise 17.1. Using the same arguments as above show that a more strong result is valid under conditions of Theorem 17.1, namely, prove that for any $n = 1, 2, \ldots$ random variables $\xi_1, \xi_2, \ldots, \xi_n$ and M_n are independent.

Now we can see that

$$E\xi_n = \frac{1}{n} \quad \text{and} \quad \text{Var}\,\xi_n = \left(\frac{1}{n} - \frac{1}{n^2}\right) = \frac{n-1}{n^2}, \quad n = 1, 2, \ldots.$$

Record indicators can be also expressed via the so-called sequential ranks of random variables $X_1, X_2, \ldots$. We say that $\rho_1, \rho_2, \ldots$ present the sequential ranks of X's, if ρ_n determines the place of X_n inside the sample $X_1, X_2, \ldots, X_n$, that is,

$$\rho_n = m, \quad m = 1, 2, \ldots, n, \quad \text{if } X_n = X_{m,n}.$$

It is known that if $X_1, X_2, \ldots$ are independent random variables having a common continuous distribution function, then ranks $\rho_1, \rho_2, \ldots$ are also independent and for any $n = 1, 2, \ldots$

$$P\{\rho_n = m\} = \frac{1}{n}, \quad m = 1, 2, \ldots, n.$$

It is evident, that events $\{\xi_n = 1\}$ and $\{\rho_n = n\}$ coincide for any $n = 1, 2, \ldots$. From here it follows that record indicators inherit the independence property of the sequential ranks and

$$P\{\xi_n = 1\} = P\{\rho_n = n\} = \frac{1}{n}, \quad n = 1, 2, \ldots.$$

Number of records in the sample $X_1, X_2, \ldots, X_n$

Let $N(n)$ denote a number of record values among random variables $X_1, X_2, \ldots, x_n$. Indeed

$$N(n) = \xi_1 + \xi_2 + \cdots + \xi_n, \quad n = 1, 2, \ldots. \tag{17.5}$$

Here $N(n)$ is presented as a sum of independent indicators. Hence

$$EN(n) = 1 + \frac{1}{2} + \cdots + \frac{1}{n} \tag{17.6}$$

and

$$\text{Var}\,(N(n)) = \sum_{k=1}^{n} \left(\frac{1}{k} - \frac{1}{k^2}\right), \quad n = 1, 2, \ldots. \tag{17.7}$$

Note that

$$EN(n) \sim \log n \tag{17.8}$$

and

$$\text{Var}\,(N(n)) \sim \log n, \quad n \to \infty. \tag{17.9}$$

If to apply to $N(n)$ classical limit theorems for sums of independent random variables it is possible to obtain the following statements which describe the asymptotic behavior of $N(n)$ under the assumption that $n \to \infty$.

Central limit theorem

$$\left| P\{N(n) - \log n < x\sqrt{\log n}\} - \Phi(x) \right| \to 0, \tag{17.10}$$

$\Phi(x) = \frac{1}{\sqrt{2\pi}} \int_{-\infty}^{x} \exp(-t^2/2)dt$ being the distribution function of the standard normal law;

Uniform estimate in Central limit theorem

$$\left| P\{N(n) - \log n < x\sqrt{\log n}\} - \Phi(x) \right| \leqslant \frac{C}{\sqrt{\log n}}, \quad n = 1, 2, \ldots, \tag{17.11}$$

C being some absolute constant.

Strong Law of Large Numbers

$$P\left\{ \lim \left(\frac{N(n)}{\log n} \right) = 1 \right\} = 1; \tag{17.12}$$

Law of Iterative Logarithm

$$P\left\{ \limsup \frac{N(n) - \log n}{(2\log n \, \log\log\log n)^{1/2}} = 1 \right\} = 1 \tag{17.13}$$

and

$$P\left\{ \liminf \frac{N(n) - \log n}{(2\log n \, \log\log\log n)^{1/2}} = -1 \right\} = 1. \tag{17.14}$$

Expression (17.5) implies that the generating function $P_n(s)$ of $N(n)$ satisfies the following equalities:

$$P_n(s) = Es^{N(n)} = Es^{N(n)} = \prod_{j=1}^{n} \left(1 + \frac{s-1}{j} \right) = \frac{s(1+s)(2+s)\cdots(n-1+s)}{n!} \tag{17.15}$$

and

$$P_n(-s) = \frac{(-1)^n s(s-1)\cdots(s-n+1)}{n!}. \tag{17.16}$$

Expression (17.16) enables us to use Stirling numbers of the first kind, which are defined by equalities

$$x(x-1)\cdots(x-n+1) = \sum_{k \geqslant 0} S_n^k x^k. \tag{17.17}$$

Exercise 17.2. Show that

$$P\{N(n) = k\} = (-1)^k \frac{S_n^k}{n!} = \left| \frac{S_n^k}{n!} \right|, \quad n = 1, 2, \ldots, \quad k = 1, 2, \ldots, n. \tag{17.18}$$

Record times

The following evident equalities tie numbers of records $N(n)$ with record times $L(n)$:

$$N(L(n)) = n, \quad n = 1, 2, \ldots, \tag{17.19}$$

$$P\{L(n) > m\} = P\{N(m) < n\}, \quad n = 1, 2, \ldots, \quad m = 1, 2, \ldots, \tag{17.20}$$

and

$$
\begin{aligned}
P\{L(n) = m\} &= P\{N(m-1) = n-1, N(m) = n\} \\
&= P\{N(m-1) = n-1, \xi_m = 1\} \\
&= P\{N(m-1) = n-1\}P\{\xi_m = 1\} \\
&= \frac{P\{N(m-1) = n-1\}}{m}, \quad 1 \leqslant n \leqslant m.
\end{aligned} \tag{17.21}
$$

Equalities (17.18) and (17.21) imply that

$$P\{L(n) = m\} = \left| \frac{S_{m-1}^{n-1}}{m!} \right|, \quad n = 1, 2, \ldots, \quad m = n, n+1, \ldots . \tag{17.22}$$

Based on properties of Stirling numbers Westcott (1977) showed that

$$P\{L(n) = m\} \sim \frac{(\log m)^{n-2}}{m^2(n-2)!} \tag{17.23}$$

as $m \to \infty$.

Relations (17.15) and (17.21) help us to find generating functions

$$Q_n(s) = Es^{L(n)}, \quad n = 1, 2, \ldots,$$

of record times. Since $P\{L(1) = 1\} = 1$, it is clear that

$$Q_1(s) = s.$$

For $n = 2, 3, \ldots, |s| < 1$ and $|z| < 1$, we get equalities

$$Q_n(s) = \sum_{m=1}^{\infty} P\{L(n) = m\}s^m = \sum_{m=1}^{\infty} \frac{1}{m} P\{N(m-1) = n-1\}s^m$$

and

$$
\begin{aligned}
Q_n(s)z^n &= \sum_{m=1}^{\infty} \frac{s^m}{m} \sum_{n=1}^{\infty} P\{N(m-1) = n-1\}z^n \\
&= z \sum_{m=1}^{\infty} \frac{s^m}{m} P_{m-1}(z) \\
&= z \sum_{m=1}^{\infty} \frac{s^m}{m} z(1+z) \cdots (m-2+z) \\
&= -\frac{z}{1-z} \sum_{m=1}^{\infty} \frac{(-s)^m}{m!} (1-z)(-z)(-1-z) \cdots (2-m-z).
\end{aligned} \tag{17.24}
$$

Note that

$$\sum_{m=0}^{\infty} \frac{(-s)^m}{m!} (1-z)(-z)(-1-z)\cdots(2-m-z) = (1-s)^{1-z}$$

$$= (1-s)\exp\{-z\log(1-s)\} \quad (17.25)$$

and then (17.24) and (17.25) imply that

$$(1-z)\sum_{n=1}^{\infty} Q_n(s)z^{n-1} = -(1-s)\exp\{-z\log(1-s)\} + 1. \quad (17.26)$$

Transforming the LHS of (17.26) in

$$\sum_{n=0}^{\infty} Q_{n+1}(s)z^n - \sum_{n=1}^{\infty} Q_n(s)z^n = s + \sum_{n=1}^{\infty} (Q_{n+1}(s) - Q_n(s))z^n,$$

we get the equality

$$1 - s - \sum_{n=1}^{\infty} (Q_{n+1}(s) - Q_n(s))z^n = (1-s)\sum_{n=0}^{\infty} \frac{(-\log(1-s))^n}{n!} z^n, \quad (17.27)$$

Let denote $R_n = Q_n(s) - Q_{n+1}(s)$, $n = 1, 2, \ldots$. Then one obtains from (17.27) that

$$R_n = (1-s)\frac{(-\log(1-s))^n}{n!}, \quad n = 1, 2, \ldots.$$

Hence

$$Q_n(s) = Q_1(s) - (R_1 + \cdots + R_{n-1})$$

$$= s - (1-s)\sum_{r=1}^{n-1} \frac{(-\log(1-s))^r}{r!}$$

$$= (1-s)\sum_{r=n}^{\infty} \frac{(-\log(1-s))^r}{r!}. \quad (17.28)$$

Exercise 17.3. Show that (17.28) can be transformed as

$$Q_n(s) = \frac{1}{(n-1)!} \int_0^{-\log(1-s)} v^{n-1} \exp(-v) dv. \quad (17.29)$$

The independence property of record indicators enables us to get joint distributions of record times.

Theorem 17.2. *For any $n = 1, 2, \ldots$ and any integers $1 = j(1) < j(2) < \cdots < j(n)$ the following equality holds:*

$$P\{L(1) = 1, L(2) = j(2), \ldots, L(n) = j(n)\}$$

$$= \frac{1}{(j(2)-1)(j(3)-1)\cdots(j(n)-1)j(n)}. \quad (17.30)$$

Proof. Evidently, the event on the left side of (17.30) coincides with the event

$$A = \{\xi_2 = 0,\ldots,\xi_{j(2)-1} = 0, \xi_{j(2)} = 1, \xi_{j(2)+1} = 0,\ldots,\xi_{j(3)-1} = 0,$$

$$\xi_{j(3)} = 1,\ldots,\xi_{j(n-1)-1} = 0, \xi_{j(n-1)} = 1, \xi_{j(n-1)+1} = 0,\ldots,\xi_{j(n)-1} = 0, \xi_{j(n)} = 1\}.$$

Due to the independence of record indicators it is possible to obtain now that

$$P\{A\} = P\{\xi_2 = 0\}\cdots P\{\xi_{j(2)-1} = 0\}P\{\xi_{j(2)} = 1\}P\{\xi_{j(2)+1} = 0\}\cdots$$

$$P\{\xi_{j(3)-1} = 0\}P\{\xi_{j(3)} = 1\}\cdots P\{\xi_{i(n-1)-1} = 0\}P\{\xi_{j(n-1)} = 1\}P\{\xi_{j(n-1)+1} = 0\}\cdots$$

$$P\{\xi_{j(n)-1} = 0\}P\{\xi_{i(n)} = 1\} = \frac{1}{j(n)}\prod_{t=2}^{n}\frac{1}{j(t)-1},$$

and this expression coincides with the LHS of (17.30)   □

Corollary 17.1. *One gets from (17.30) that*

$$P\{L(n) = m\} = \sum_{1 < j(2) < \cdots < j(n-1) < m} \frac{1}{(j(2)-1)(j(3)-1)\cdots(j(n-1)-1)(m-1)m},$$

$$m = n, n+1,\ldots. \quad (17.31)$$

In particular,

$$P\{L(2) = m\} = \frac{1}{(m-1)m}, \quad m = 2, 3,\ldots, \quad (17.32)$$

and hence

$$P\{L(2) > m\} = \frac{1}{m}, \quad m = 1, 2,\ldots \quad (17.33)$$

Exercise 17.4. Show that the following equalities are valid for the conditional distributions of record times:

$$P\{L(n+1) = m \mid L(n) = j(n), L(n-1) = j(n-1),\ldots, L(2) = j(2), L(1) = 1\}$$

$$= \frac{j(n)}{m(m-1)}, \quad (17.34)$$

$$1 = j(1) < j(2) < \cdots < j(n-1) < j(n) < m,$$

and

$$P\{L(n+1) = m \mid L(n) = j\} = \frac{j}{m(m-1)}, \quad n \leqslant j < m. \quad (17.35)$$

Remark 17.1. It follows from equalities (17.34) and (17.35) that the sequence of record times $L(1)$, $L(2)$,... presents a Markov chain.

Williams (1973) proved that the following representation for record times is valid:

$$L(1) = 1, \ L(n+1) = [L(n)\exp(Z_n)]+1, \quad n = 1, 2, \ldots, \tag{17.36}$$

where Z_1, Z_2,... are independent random variables having the standard exponential distribution and $[x]$ denotes the entire part of x.

Exercise 17.5. Show that (17.36) can be rewritten in the following alternative form:

$$L(1) = 1, \ L(n+1) = \left[\frac{L(n)}{U_n}\right]+1, \quad n = 1, 2, \ldots, \tag{17.37}$$

where U_1, U_2,... are independent random variables with the common uniform $U([0,1])$ distribution.

Some problems appear with moments of record times. It follows from (17.32) that $EL(2) = \infty$. Hence $EL(n) = \infty$ for any $n = 2, 3, \ldots$, and $E(L(n))^\alpha = \infty$, $n = 2, 3, \ldots$, if $\alpha \geqslant 1$.

In this situation the following expression for logarithmic moment may be useful:

$$E\log L(n) = n - C - 2^{-(n+1)} + O(3^{-n}), \quad n \to \infty, \tag{17.38}$$

where $C = 0.5772\ldots$ is Euler's constant.

Note also that the next analogous expression is valid for moments $E(L(n))^{1-\beta}$, $\beta > 0$:

$$E(L(n))^{1-\beta} = \frac{1}{\Gamma(\beta)}\left\{\beta^{-n} + \frac{\beta-1}{2}(\beta+1)^{-n} + O((\beta+2)^{-n})\right\}, \quad n \to \infty. \tag{17.39}$$

It was mentioned above that the relation

$$P\{L(n) > m\} = P\{N(m) < n\}, \quad n = 1, 2, \ldots, \ m = 1, 2, \ldots$$

ties the distributions of $N(n)$ and $L(n)$. Due to this equality the most part of the limit theorems given for $N(n)$, were overworked (see, for example, Renyi (1962)) into the corresponding theorems for record times $L(n)$ (exactly saying, for $\log L(n)$) as follows:

a) *Central limit theorem*

$$\sup_x \left|P\left\{\log L(n) - n < x\sqrt{n}\right\} - \Phi(x)\right| \to 0, \quad n \to \infty. \tag{17.40}$$

b) *Strong Law of Large Numbers*

$$P\left\{\lim\log\frac{L(n)}{n} = 1\right\} = 1, \quad n \to \infty. \tag{17.41}$$

c) *Law of Iterative Logarithm*

$$P\left\{\limsup \frac{\log L(n) - n}{(2n \log\log n)^{1/2}} = 1\right\} = 1, \quad n \to \infty, \tag{17.42}$$

and

$$P\left\{\liminf \frac{\log L(n) - n}{(2n \log\log n)^{1/2}} = -1\right\} = 1, \quad n \to \infty. \tag{17.43}$$

Record values

Record values $X(1) < X(2) < \cdots < X(n) < \cdots$ are expressed via record times $L(1) < L(2) < \cdots < L(n) < \cdots$ as follows:

$$X(n) = M_{L(n)} = X_{L(n),L(n)}, \quad n = 1, 2, \ldots \tag{17.44}$$

Thus,

$$P\{X(n) < x\} = P\{M_{L(n)} < x\} = \sum_{m=n}^{\infty} P\{M_{L(n)} < x \mid L(n) = m\}P\{L(n) = m\}$$

$$= \sum_{m=n}^{\infty} P\{M_m < x \mid L(n) = m\}P\{L(n) = m\}. \tag{17.45}$$

Since

$$\{L(n) = m\} = \{\xi_1 + \xi_2 + \cdots + \xi_{m-1} = n - 1, \xi_m = 1\},$$

and the vector of record indicators $(\xi_1, \xi_2, \ldots, \xi_{m-1}, \xi_m)$ and maxima M_m are independent (see the result of exercise 17.1) it follows that for any $m = 1, 2, \ldots$, the event $\{L(n) = m\}$ and the random variable M_m are independent. Hence the RHS of (17.45) can be transformed as follows:

$$\sum_{m=n}^{\infty} P\{M_m < x \mid L(n) = m\}P\{L(n) = m\} = \sum_{m=n}^{\infty} P\{M_m < x\}P\{L(n) = m\}$$

$$= \sum_{m=n}^{\infty} F^m(x)P\{L(n) = m\}$$

$$= E(F(x))^{L(n)}. \tag{17.46}$$

Finally we get the relation

$$P\{X(n) < x\} = E(F(x))^{L(n)} = Q_n(F(x)), \tag{17.47}$$

where the corresponding expressions for the generating function $Q_n(s)$ is given in (17.29). Thus,

$$P\{X(n) < x\} = \frac{1}{(n-1)!} \int_0^{-\log(1-F(x))} v^{n-1} \exp(-v)\,dv, \quad -\infty < x < \infty,$$

$$n = 1, 2, \ldots. \tag{17.48}$$

Equality (17.48) presents a general form for distribution functions of record values in the situation when $F(x)$ is a continuous distribution function. Considering the partial case of (17.48) for $F(x) = 1 - \exp(-x)$, $x > 0$, that is, for the case, when initial X's have the standard $E(1)$ exponential distribution, one gets that

$$P\{X(n) < x\} = \frac{1}{(n-1)!} \int_0^x v^{n-1} \exp(-v) dv, \quad 0 \leqslant x < \infty, \ n = 1, 2, \ldots. \qquad (17.49)$$

Thus, in this situation $X(n)$ has the Gamma distribution with the parameter n and the following representation holds for $X(n)$:

$$X(n) \overset{d}{=} X_1 + \cdots + X_n, \quad n = 1, 2, \ldots. \qquad (17.50)$$

Moreover, it can be proved that if $X_1, X_2, \ldots$ are independent random variables having the standard $E(1)$ exponential distribution, then (17.50) can be rewritten in a more strong form:

$$\{X(1), X(2), \ldots, X(n)\} \overset{d}{=} \{S_1, S_2, \ldots, S_n\}, \quad n = 1, 2, \ldots, \qquad (17.51)$$

where $S_k = X_1 + \cdots + X_k$, $k = 1, 2, \ldots$.

It follows from (17.51) that in this case random variables $X(1), X(2) - X(1), \ldots, X(n) - X(n-1), \ldots$ are independent and each of them has the standard exponential $E(1)$ distribution.

Are there such kind of simple representations as given in (17.51) for other distributions of X's? It is well known that if a random variable X has a continuous distribution function F, then the transformation $Y = F(X)$ produces the uniformly $U([0,1])$ distributed random variable Y. Since this transformation does not change the order of X's, the vector $\{U(1), U(2), \ldots, U(n)\}$ of records in a sequence $U_1, U_2, \ldots, U_n$, where U's are independent $U([0,1])$ distributed random variables, coincides in distribution with the vector $\{F(X(1)), F(X(2)), \ldots, F(X(n))\}$.

Let F be a continuous distribution function and G is the inverse of F. Then the following equality is also valid for any $n = 1, 2, \ldots$:

$$\{X(1), X(2), \ldots, X(n)\} \overset{d}{=} \{G(U(1)), G(U(2)), \ldots, G(U(n))\}. \qquad (17.52)$$

Now let record values $X(1) < X(2) < \cdots$ and $Y(1) < Y(2) < \cdots$ correspond to X's and Y's with continuous distribution functions F and H respectively. Then (17.52) implies that

$$\{X(1), X(2), \ldots, X(n)\} \overset{d}{=} \{G(H(Y(1))), G(H(Y(2))), \ldots, G(H(Y(n)))\},$$

$$n = 1, 2, \ldots, \qquad (17.53)$$

where G is the inverse function of F. Combining (17.51) and (17.53) we come to the next equality:

$$\{X(1), X(2), \ldots, X(n)\} \overset{d}{=} \{H(Z_1), H(Z_1 + Z_2), \ldots, H(Z_1 + Z_2 + \cdots + Z_n)\},$$

$$n = 1, 2, \ldots, \quad (17.54)$$

where $H(x) = G(1 - \exp(-x))$, G is the inverse of F and $Z_1, Z_2, \ldots, Z_n$ are independent exponentially $E(1)$ distributed random variables.

The joint distribution functions of record values have a rather complicate form. Say, in the simplest case ($n = 2$) the following equalities hold:

$$P\{X(1) < x_1, X(2) < x_2\} = \sum_{m=2}^{\infty} P\{X_1 < x_1, \max\{X_2, \ldots, X_{m-1}\} \leqslant X_1, X_1 < X_m < x_2\}$$

$$= \sum_{m=2}^{\infty} \int_{-\infty}^{\min(x_1, x_2)} F^{m-2}(u)(F(x_2) - F(u))dF(u)$$

$$= \int_{-\infty}^{\min(x_1, x_2)} \frac{F(x_2) - F(u)}{1 - F(u)} dF(u)$$

$$= \int_{0}^{F(\min(x_1, x_2))} \frac{F(x_2) - u}{1 - u} du$$

$$= (1 - F(x_2)) \ln(1 - F(\min(x_1, x_2)) + F(\min(x_1, x_2)).$$

It means that

$$P\{X(1) < x_1, X(2) < x_2\} = \begin{cases} (1 - F(x_2)) \ln(1 - F(x_1)) + F(x_1), & \text{if } x_1 < x_2, \\ (1 - F(x_2)) \ln(1 - F(x_2)) + F(x_2), & \text{otherwise.} \end{cases} \quad (17.55)$$

The general expression ($n = 2, 3, \ldots$) for the joint distribution functions of the record values corresponding to any continuous distribution function F is given as

$$P\{X(1) < x_1, X(2) < x_2, \ldots, X(n) < x_n\} = \int \cdots \int \prod_{j=1}^{n-1} \frac{dF(u_j)}{1 - F(u_j)} dF(u_n), \quad (17.56)$$

where integration on the RHS of (17.56) holds over the set

$$B = \{u_j < x_j, \ j = 1, 2, \ldots, n, \ -\infty < u_1 < \cdots < u_n < \infty\}.$$

A more simple expression is valid for the joint density functions of the record values.

Suppose that F is an absolutely continuous disrtibution function with a density function f. Let us denote

$$R(x) = \frac{f(x)}{1 - F(x)}.$$

Then the joint density function of record values $X(1), X(2), \ldots, X(n)$ is given as

$$f_{1,2,\ldots,n}(x_1, x_2, \ldots, x_n) = \begin{cases} R(x_1)R(x_2)\cdots R(x_{n-1})f(x_n), & \text{if } X_1 < X_2 < \cdots < X_n; \\ 0, & \text{otherwise.} \end{cases}$$

(17.57)

Exercise 17.6. Write the joint density function $f_{1,2,\ldots,n}$ for the cases, when $X \sim E(1)$ and $X \sim U([0,1])$.

Since $X(n) = \max\{X_1, \ldots, X_{L(n)}\}$, the limit distributions of record values must be close to the analogous distributions of maximal order statistics. As it is known (see chapter 11) there are three types of asymptotic distributions for the suitable centering and normalizing maxima

$$\frac{M_n - b(n)}{a(n)}.$$

They are:

$$\Lambda(x) = \exp(-\exp(-x)),$$

$$\Phi_\alpha(x) = \begin{cases} 0, & \text{if } x < 0, \\ \exp(-x^{-\alpha}), & \text{if } x > 0 \end{cases}$$

and

$$\Psi_\alpha(x) = \begin{cases} \exp(-(-x)^\alpha), & \text{if } x < 0, \\ 1, & \text{if } x > 0, \end{cases}$$

Indeed, under the corresponding random centering and normalizing we will obtain the same limit distributions for random variables

$$\frac{X(n) - b(L(n))}{a(L(n))}.$$

The more interesting question arises: what types of asymptotic distributions can we get for nonrandomly normalized record values? What types of the limit distribution functions $T(x)$, where

$$T(x) = \lim P\{X(n) - B(n) < xA(n)\}, \quad n \to \infty,$$

can be obtained under the corresponding choice of the normalizing and centering constants $A(n) > 0$ and $B(n)$?

Tata (1969) (see also Resnick (1973)) found that all possible limits $T(x)$ have the form

$$T(x) = \Phi(g_k(x)), \quad k = 1, 2, 3,$$

(17.58)

where

$$\Phi(x) = (2\pi)^{-1/2} \int_{-\infty}^{x} \exp\left(-\frac{t^2}{2}\right) dt$$

and $g_k(x)$, $k = 1, 2, 3$, are determined (up to linear transformations) as follows:

$$g_1(x) = x;$$

$$g_2(x) = \begin{cases} \gamma \log x, \gamma > 0, & \text{if } x > 0, \\ -\infty, & \text{if } x < 0; \end{cases}$$

$$g_3(x) = \begin{cases} -\gamma \log(-x), \gamma > 0, & \text{if } x < 0, \\ \infty, & \text{if } x > 0. \end{cases}$$

Exercise 17.7. Taking into account representation (17.5) find the centering and normalizing constants as well as the corresponding limit distribution for the record values $X(n)$ in the situation, when X's have the standard $E(1)$ exponential distribution.

Above there were given some main results for the upper records generated by a sequence of independent random variables $X_1, X_2, \ldots$ having a common continuous distribution function. All of these results can be reformulated easily for the lower records (X_n is a lower record value if $X_n < \min\{X_1, X_2, \ldots, X_{n-1}\}$). For this purpose enough instead of X's take random variables $Y_k = -X_k$, $k = 1, 2, \ldots$. A lot of other record schemes were suggested for the so-called weak records, the k-th records, records in the sequences of discrete random variables, records generated by different types of dependent X's and so on.

More intensive investigation of records and a lot of results for record times and record values for different record models can be found in monographs Ahsanullah (1995), Arnold, Balakrishnan and Nagaraja (1998) and Nevzorov (2000, 2001).

Check your solutions

Exercise 17.1 (hint). The proof is analogous to the proof of Theorem 17.1. It is enough to show that

$$P\{\xi_1 = 1, \ \xi_2 = 1, \ldots, \ \xi_n = 1, \ M_n < x\} = \frac{F^n(x)}{n!}, \quad n = 2, 3, \ldots$$

Exercise 17.2 (hint). Apply equalities (17.15)–(17.17).

Exercise 17.3 (hint). It is enough to use the following well-known identity for Poisson probabilities:

$$\sum_{r=n}^{\infty} \frac{\lambda^r e^{-\lambda}}{r!} = \frac{1}{(n-1)!} \int_0^{\lambda} v^{n-1} e^{-v} dv, \quad n = 1, 2, \ldots, \lambda > 0.$$

Exercise 17.4 (hint). These results follow immediately from equality (17.30).

Exercise 17.5 (hint). To prove presentation (17.37) it is enough to recall the fact that if $Z \sim E(1)$ and $X \sim U([0,1])$ then

$$1 - \exp(-Z) \overset{d}{=} U$$

and hence

$$\exp(-Z) \overset{d}{=} \frac{1}{1-U} \overset{d}{=} \frac{1}{U}.$$

Exercise 17.6 (hint). Use relation (17.57) with $R(x) = 1$ and $f(x) = \exp(-x)$, $x \geqslant 0$, when $X \sim E(1)$, and use this relation with $R(x) = 1/(1-x)$ and $f(x) = 1$, $0 < x < 1$, when $X \sim U([0,1])$.

Exercise 17.7 (answers). If $F(x) = 1 - \exp(-x)$, $x \geqslant 0$, then

$$\left| P\{X(n) - n < x\sqrt{n}\} - \Phi(x) \right| \to 0, \quad n \to \infty,$$

where $\Phi(x) = (2\pi)^{-1/2} \int_{-\infty}^{x} \exp(-t^2/2) dt$.

Chapter 18

Characterizations of Distributions Based on Properties of Order Statistics

Some properties of order statistics which characterized the concrete probability distributions are investigated below.

Consider a sequence of independent random variables $X_1, X_2, \ldots$ having a common distribution function F and the corresponding sets of order statistics $X_{1,n} \leqslant \cdots \leqslant X_{n,n}$, $n = 1, 2, \ldots$. There are a lot of characterizations of distribution functions F based on different properties of order statistics and their moment characteristics. The history of such characterizations numbers more than 60 years. As the first ones in this direction we can mention the results of Moriguti (1951, 1953), Fisz (1958), Ferguson (1967), Rossberg (1972) and so on (for details see, for example, Ahsanullah and Nevzorov (2001), Arnold, Balakrishnan and Nagaraja (1992)). It is interesting to mention that the most part of characterizations were obtained for exponential and uniform families of distributions. The most part of such characterizing theorem are discussed in Kamps (1991) and in Rao and Shanbhag (1998, pp. 231–256). For example, one of the classical result (given by Desu (1977)) for the exponential family of distributions is based on the following equalities in distribution:

$$nX_{1,n} \overset{d}{=} X_1. \tag{18.1}$$

Proof. Some discussions and the further generalizations of this result can be found in Arnold, Balakrishnan and Nagaraja (1992). We mention here one more classical result given by Ahsanullah (1989), who showed that under some additional restrictions on distribution function f the condition that order statistics $X_{n-1,n}$ and the sample range $X_{n,n} - X_{1,n}$ are identically distributed for some $n > 1$ characterizes the uniform family of distributions.

Hence we will try to give below a series of some characterizations of nontraditional distributions which were obtained at the last time. $\qquad\square$

1. Let us begin with the following model. Below $U \sim U([0,1])$ is a random variable having the uniform on the interval $[0,1]$ distribution and U is independent of X's. It is

M. Ahsanullah et al., *An Introduction to Order Statistics*,
Atlantis Studies in Probability and Statistics 3,
DOI: 10.2991/978-94-91216-83-1_18, © Atlantis Press 2013

interesting to find distributions such that equality (18.2) holds in distribution:

$$X_{k,k}U \overset{d}{=} X_{m,m}, \qquad (18.2)$$

where $1 \leqslant m < k, X_{k,k} = \max\{X_1, X_2, \dots, X_k\}$ and $X_{m,m} = \max\{X_1, X_2, \dots, X_m\}$.

The following result is valid.

Theorem 18.1. *Let $X_1, X_2, \dots$ be independent nonnegative random variables having a common continuous distribution function F and $U \sim U([0,1])$ is independent of X's. Then equality (18.2) holds if and only if*

$$F(x) = H(cx), \qquad (18.3)$$

where c is any positive constant and

$$H(x) = (1 + x^{(m-k)/m})^{1/(m-k)}, \quad 0 < x < \infty. \qquad (18.4)$$

Proof. Since $P\{X_{k,k} < x\} = (F(x))^k$ and $P\{X_{m,m} < x\} = (F(x))^m$, it follows from (18.2) that

$$\int_0^1 F^k\left(\frac{x}{u}\right) du = F^m(x)$$

and then

$$\int_x^\infty \frac{F^k(v)}{v^2} dv = \frac{F^m(x)}{x}. \qquad (18.5)$$

Under conditions of the theorem F is a continuous function. Then it follows from (18.5) that F has a continuous derivative and the following equality can be obtained after differentiating the both sides of (18.5):

$$\frac{F^k(x)}{x^2} = \frac{F^m(x)}{x^2} - \frac{mF^{m-1}(x)F'(x)}{x}. \qquad (18.6)$$

Equality (18.6) is rewritten as

$$F(x) - F^{k-m+1}(x) = mxF'(x). \qquad (18.7)$$

Let $G(x) = F^{-1}(x)$ be the inverse function to $F(x)$. Since

$$F(G(x)) = x, \quad 0 < x < 1,$$

and

$$F'(G(x)) = \frac{1}{G'(x)}$$

it follows from (18.7) that

$$x - x^{k-m+1} = \frac{mG(x)}{G'(x)}, \quad 0 < x < 1,$$

and hence

$$(\log G(x))' = \frac{m}{x - x^{k-m+1}}. \tag{18.8}$$

One gets from (18.8) that

$$G(x) = \frac{cx^m}{(1 - x^{k-m}))^{m/(k-m)}}, \quad 0 < x < 1, \tag{18.9}$$

where c is any positive constant. Then

$$F(x) = \frac{x^{1/m}}{(c_1 + x^{(k-m)/m})^{1/(k-m)}}, \quad 0 < x < \infty, \tag{18.10}$$

where c_1 also is any positive constant. It is not difficult to see that the statement of the theorem follows from (18.10). □

Remark 18.1. If $k = 2$ and $m = 1$, then (18.10) gives the following form of $F(x)$:

$$F(x) = H(cx),$$

where c is any positive constant,

$$H(x) = \begin{cases} \dfrac{x}{1+x}, & \text{if } 0 < x < \infty, \\ 0, & \text{if } x \leqslant 0. \end{cases}$$

Exercise 18.1. It is not difficult to obtain some natural generalizations of Theorem 18.1. For example, consider the following situation. Let now $U_\alpha \sim \mathrm{Pow}(1, \alpha)$ be the random variable having the power function distribution with distribution function

$$R_\alpha(x) = x^\alpha, \quad 0 \leqslant x \leqslant 1,$$

where $\alpha > 0$. Indeed, if $\alpha = 1$, then $U_1 \sim U([0,1])$ has the standard uniform distribution. Let us consider the following equality:

$$X_{k,k} U_\alpha \overset{d}{=} X_{m,m}, \quad 1 \leqslant m < k. \tag{18.11}$$

Show that relation (18.11) holds if and only if

$$F(x) = H_\alpha(cx),$$

where c is any positive constant,

$$H_\alpha(x) = \begin{cases} x^{\alpha/m}(1 + x^{\alpha(m-k)/k})^{1/(m-k)}, & \text{if } 0 < x < \infty, \\ 0, & \text{if } x \leqslant 0. \end{cases} \tag{18.12}$$

Comment. Indeed the result given as *Exercise 18.1* can considered as some generalization of Theorem 18.1, which deals with the partial case ($\alpha = 1$) of equality (18.11), but really the short proof of this general relation can be based on the statement of Theorem 18.1. Really, it is enough to mention that

$$U_\alpha \stackrel{d}{=} U^{1/\alpha},$$

where $U \sim U([0,1])$, and then (18.11) is equivalent to equality

$$X_{k,k}^\alpha U \stackrel{d}{=} X_{m,m}^\alpha.$$

The next steps to obtain expression (18.12) for distribution function $H_\alpha(x)$ are evident.

It is interesting to compare relation (18.2) with equivalent (at least, from the first sight) equality

$$X_{k,k} \stackrel{d}{=} \frac{X_{m,m}}{U}, \quad 1 \leqslant m < k. \tag{18.13}$$

The following result is valid.

Theorem 18.2. *Let $X_1, X_2, \ldots$ be independent nonnegative random variables having a common continuous distribution function F and let $U \sim U([0,1])$ be independent of X's. Then equality* (18.13) *holds if and only if*

$$F(x) = H(cx),$$

where c is any positive constant,

$$H(x) = \begin{cases} (1 - x^{(m-k)/k})^{1/(k-m)}, & \text{if } x > 1, \\ 0, & \text{if } x \leqslant 1. \end{cases} \tag{18.14}$$

Proof. Equality (18.13) implies that

$$F^k(x) = \int_0^x F^m(ux)du = \int_0^x F^m(v)dv/x$$

and hence

$$xF^k(x) = \int_0^x F^m(v)dv. \tag{18.15}$$

Since $F(x)$ is continuous one can differentiate the RHS of (18.15). It means that the LHS of (18.15) is also differentiable. Then one gets that

$$F^k(x) + kxF^{k-1}(x)F'(x) = F^m(x)$$

and hence

$$F^{k-m}(x) + kxF^{k-m-1}(x)F'(x) = 1. \tag{18.16}$$

Taking into account equation

$$\frac{1}{kx} = \frac{F'(x)F^{k-m-1}(x)}{1 - F^{k-m}(x)},$$

one gets that (18.16) holds if the following equality is valid:

$$c_1 x^{1/k} = (1 - F^{k-m}(x))^{1/(m-k)}, \tag{18.17}$$

Now the statement of Theorem 18.2 is evident.   □

Exercise 18.2. Now let us consider a more general equation than (18.13):

$$X_{k,k} \stackrel{d}{=} \frac{X_{m,m}}{U_\alpha}, \quad 1 \leqslant m < k, \tag{18.18}$$

where $U_\alpha \sim \text{Pow}(1, \alpha)$, $\alpha > 0$.

Show that (18.18) holds if and only if

$$F(x) = H_\alpha(cx),$$

where c is any positive constant,

$$H_\alpha(x) = \begin{cases} (1 - x^{\alpha(m-k)/k})^{1/(k-m)}, & \text{if } x > 1, \\ 0, & \text{if } x \leqslant 1. \end{cases} \tag{18.19}$$

Remark 18.2. It is interesting to see that similar equations (18.2) and (18.13) (as well as equalities (18.11) and (18.18)) characterize different families of distributions.

Exercise 18.3. One can see that (18.13) is equivalent to equality

$$\frac{U}{X_{m,m}} \stackrel{d}{=} \frac{1}{X_{k,k}}, \quad 1 \leqslant m < k. \tag{18.20}$$

Let us consider now random variables $Y_j = 1/X_j$, $j = 1, 2, \ldots$, instead of X's. It allows us to rewrite (18.20) as

$$Y_{1,m} U \stackrel{d}{=} Y_{1,k}. \tag{18.21}$$

Let $Y_1, Y_2, \ldots$ be independent identically distributed positive random variables,

$$Y_{1,m} = \min\{Y_1, Y_2, \ldots, Y_m\}, \quad Y_{1,k} = \min\{Y_1, Y_2, \ldots, Y_k\}, \quad 1 \leqslant m < k.$$

Find the family of distributions which are characterized by relation (18.21).

2. The next model is very close to the previous scheme.

Let us return to equality (18.2)

$$X_{k,k}U \overset{d}{=} X_{m,m},$$

in the situation, when X's are positive random variables. One can see that (18.2) is equivalent to the following equality:

$$\log X_{k,k} + \log U \overset{d}{=} \log X_{m,m}, \quad 1 \leqslant m < k. \tag{18.22}$$

Consider now random variables $Y_j = \log X_j + \mu$, where μ is any fixed constant. Taking into account that $Z = -\log U$, where $U \sim U([0,1])$, has the standard $E(1)$ exponential distribution with the distribution function $T(x) = 1 - \exp(-x)$, $x > 0$, one can rewrite (18.22) as

$$Y_{k,k} - Z \overset{d}{=} Y_{m,m}, \quad 1 \leqslant m < k. \tag{18.23}$$

Thus, equality (18.23) holds for random variables $Y_1, Y_2, \ldots$ if and only if equation (18.2) is valid for X's such that

$$X_j = \exp(Y_j - \mu), \quad j = 1, 2, \ldots.$$

Note that

$$P\{Y_j < x\} = P\{X < \exp(x - \mu)\}. \tag{18.24}$$

Combining these details and the statement of Theorem 18.1 one gets the following result.

Theorem 18.3. *Let $Y_1, Y_2, \ldots$ be independent random variables having a common continuous distribution function F and let $Z \sim E(1)$ be independent of Y's. Then equality (18.23) holds if and only if*

$$F(x) = H(c(x - \mu)),$$

where $-\infty < \mu < \infty$ and $c > 0$ are any constants and

$$H(x) = \left(1 + \exp\left\{-\frac{k-m}{m}x\right\}\right)^{1/(m-k)}, \quad -\infty < x < \infty. \tag{18.25}$$

Corollary 18.1. *The simplest partial case of (18.23) with $k = 2$ and $m = 1$ deals with equality*

$$Y_1 \overset{d}{=} Y_{2,2} - Z. \tag{18.26}$$

It follows from the result of Theorem 18.3 that (18.26) is valid if and only if

$$F(x) = \frac{1}{1 + \exp\{-c(x-\mu)\}}, \quad -\infty < x < \infty, \tag{18.27}$$

where $-\infty < \mu < \infty$ and $c > 0$ are any constants.

The following result (Theorem 1.4 by M. Ahsanullah, G. Yanev and O. Constantin (2012)) is more general than statement of the above corollary.

Theorem 18.4. *Let Y_1 and Y_2 be independent random variables with distribution functions $F_1(x) = F^\alpha(x)$ and $F_2(x) = F^\beta(x)$ correspondingly, where F is some continuous distribution function, $\alpha > 0$, $\beta > 0$ and let $Z \sim E(1)$ be independent of Y's. Then equality*

$$Y_1 \overset{d}{=} \max\{Y_1, Y_2\} - Z$$

holds if and only if

$$F(x) = \left(1 + \exp\left\{-\frac{\beta(x-c)}{\alpha}\right\}\right)^{-1/\beta}, \quad -\infty < x < \infty,$$

where c is an arbitrary constant.

Exercise 18.4. Now instead of (18.23) consider the following equality:

$$Y_{k,k} \overset{d}{=} Y_{m,m} + Z, \quad 1 \leqslant m < k, \tag{18.28}$$

where again $Z \sim E(1)$ is independent of X's. Show that (18.28) holds if and only if

$$P\{Y < x\} = H(\exp\{c(x-\mu)\}),$$

where $c > 0$ and $-\infty < \mu < \infty$ are any constants,

$$H(x) = \begin{cases} (1 - x^{(m-k)/k})^{1/(k-m)}, & \text{if } x > 1, \\ 0, & \text{otherwise.} \end{cases} \tag{18.29}$$

In particular, it follows from (18.29) that equality

$$Y_{2,2} \overset{d}{=} Y_1 + Z \tag{18.30}$$

holds if and only if

$$P\{Y < x\} = (1 - \exp(-c(x-\mu))), \quad x > \mu,$$

where $c > 0$ and $-\infty < \mu < \infty$ are any constants. It means that equation (18.30) characterizes the family of the exponential distributions.

3. Now we consider one more scheme connected with characterizations of distributions by some regressional properties of order statistics. Such type of investigations of properties of order statistics and record values became very popular at the last time.Below we will consider one of the simplest situations and will give the characterization of distributions based on one regressional property of midrange $M = (X_{1,3} + X_{3,3})/2$. The following result is valid.

Theorem 18.5. *Let X_1, X_2, X_3 be independent identically distributed random variables having a finite expectation and having a continuous distribution function $F(x)$ and a probability density function $f(x)$. Let also $f(x) > 0$, if $\gamma < x < \delta$, where $\gamma = \inf\{x : F(x) > 0\}$ and $\delta = \sup\{x : F(x) < 1\}$.*

Then the relation

$$E(M \mid X_{2,3} = x) = x \text{ a.s.} \tag{18.31}$$

holds if and only if

$$F(x) = F_2\left(\frac{x-\mu}{\sigma}\right),$$

where $-\infty < \mu < \infty$, $\sigma > 0$ are any constants and

$$F_2(x) = \frac{1}{2}\left(1 + \frac{x}{(2+x^2)^{1/2}}\right), \quad -\infty < x < \infty, \tag{18.32}$$

is the distribution function of the Student's t_2-distribution.

Proof. Note that equality (18.31) is equivalent to relation

$$E((X_1 + X_2 + X_3) \mid X_{2,3} = x) = 3x$$

and hence to the relation

$$E(X_1 \mid X_{2,3} = x) = x. \tag{18.33}$$

The next general formulae for order statistics (see, for example, Nagaraja, Nevzorov (1977)) is valid:

$$E(X_1 \mid X_{k,n} = x) = \frac{x}{n} + \frac{k-1}{n}E(X \mid X < x) + \frac{n-k}{n}E(X \mid X > x). \tag{18.34}$$

In our case using (18.34), when $n = 3$ and $k = 2$, we obtain from (18.33) that

$$x = \frac{1}{3}\left(x + \frac{1}{F(x)}\int_\gamma^x tf(t)dt + \frac{1}{1-F(x)}\int_x^\delta tf(t)dt\right), \quad \gamma < x < \delta. \tag{18.35}$$

Since $E|X| < \infty$, then

$$\lim_{x\to\gamma} xF(x) = \lim_{x\to\delta} x(1 - F(x)) = 0$$

and hence

$$\frac{1}{F(x)} \int_\gamma^x tf(t)dt = x - \frac{1}{F(x)} \int_\gamma^x F(t)dt \tag{18.36}$$

and

$$\frac{1}{1-F(x)} \int_x^\delta tf(t)dt = x + \frac{1}{1-F(x)} \int_x^\delta (1-F(t))dt. \tag{18.37}$$

One gets from (18.35)–(18.37) that

$$F(x) \int_x^\delta (1-F(t))dt = (1-F(x)) \int_\gamma^x F(t)dt, \quad \gamma < x < \delta, \tag{18.38}$$

and hence

$$\frac{F(x)}{\int_\gamma^x F(t)dt} = \frac{1-F(x)}{\int_x^\delta (1-F(t))dt}, \quad \gamma < x < \delta. \tag{18.39}$$

It means that

$$\left(\log \int_\gamma^x F(t)dt \right)' = - \left(\log \int_x^\delta (1-F(t))dt \right)', \quad \gamma < x < \delta, \tag{18.40}$$

and

$$\int_\gamma^x F(t)dt = \frac{c}{\int_x^\delta (1-F(t))dt}, \quad \gamma < x < \delta, \tag{18.41}$$

where c is any positive constant. It follows from (18.41) that

$$F(x) = \frac{c(1-F(x))}{\left(\int_x^\delta (1-F(t))dt \right)^2}$$

and

$$\int_x^\delta (1-F(t))dt = c_1 \left(\frac{1-F(x)}{F(x)} \right)^{1/2}, \tag{18.42}$$

where $c_1 = c^{1/2} > 0$. After one more differentiating one obtains that the following equality

$$(F(x)(1-F(x)))^{3/2} = c_2 f(x), \quad \gamma < x < \delta, \tag{18.43}$$

holds with some positive constant c_2. Let $G(x), 0 \leqslant x \leqslant 1$, be the inverse function for $F(x)$, that is

$$F(G(x)) = x, \quad 0 < x < 1.$$

Under conditions of the theorem $G(0) = \gamma$ and $G(1) = \delta$.
It is possible to rewrite (18.43) as

$$c_2 f(G(x)) = (x(1-x))^{3/2}, \quad 0 < x < 1. \tag{18.44}$$

Since $G'(x) = 1/f(G(x))$ it follows from (18.44) that

$$G'(x) = c_2/(x(1-x))^{3/2}$$

and

$$G(x) = c_2 \int_{1/2}^{x} (u(1-u))^{-3/2} du + \mu, \qquad (18.45)$$

where μ is any constant. One gets from (18.45) that $\gamma = G(0) = -\infty$ and $\delta = G(1) = \infty$.
Note that

$$G_1(x) = \int_{1/2}^{x} (u(1-u))^{-3/2} du = \frac{2x-1}{2(x-x^2)^{1/2}}, \quad 0 < x < 1,$$

is the inverse function to distribution function

$$H(x) = \frac{1}{2} \left(1 + \frac{x}{(1+x^2)^{1/2}} \right), \quad -\infty < x < \infty.$$

Then one gets that $F(x)$, which is inverse to $G(x)$, has the following form:

$$F(x) = H \left(\frac{x-\mu}{c_2} \right) = \frac{1}{2} \left(1 + \frac{x-\mu}{(c_2^2 + (x-\mu)^2)^{1/2}} \right), \qquad (18.46)$$

where $c_2 > 0$ and $-\infty < \mu < \infty$. It is enough now to mention only that family of distributions given by (18.46) coincides with the corresponding family Student's t_2-distributions.
Thus, Theorem 18.5 is proved. □

Exercise 18.5. Consider now one more regressional scheme, which is rather close to previous one and there we have one more characterization of Student's t_2-distributions. Prove the next theorem.

Theorem 18.6. *Let X_1, X_2, X_3 be independent identically distributed random variables with the zero expectation and a continuous distribution function $F(x)$. Then equality*

$$E(X_{1,3} X_{3,3} \mid X_{2,3} = x) = c, \qquad (18.47)$$

where $c < 0$, holds if and only if

$$F(x) = H(x/\sigma),$$

where σ is any positive constant and

$$H(x) = \frac{1}{2} \left(1 + \frac{x}{(2+x^2)^{1/2}} \right), \quad -\infty < x < \infty.$$

Remark 18.3. Indeed, the result of Theorem 18.6 gives one more characterization of the family of Student's t_2-distributions by regressional properties of order statistics.

Check your solutions

Exercise 18.1 (hint). The proof of this result is analogous to the proof of Theorem 18.1.

Exercise 18.2 (hint). Take into account the arguments which were suggested in Comment 18.1.

Exercise 18.3 (hint). Let the equality

$$Y_{1,m}U \stackrel{d}{=} Y_{1,k}.$$

is valid. Take now new random variables $V_r = -Y_r$, $r = 1, 2, \ldots, k$. One gets that

$$V_{m,m}U \stackrel{d}{=} V_{k,k},$$

that is equality (18.2) is valid (with changing k by m and m by k). Hence it is possible to apply the result of Theorem 18.1. Thus we get the distribution of V's and then it gives us immediately the distribution of Y's.

Exercise 18.4 (hint). Take into account the ideas given above and the result of Theorem 18.2

Exercise 18.5 (hint). To prove theorem 18.6 it is necessary to remember, that order statistics $X_{1,3}$ and $X_{3,3}$ are conditionally independent under any fixed value of $X_{2,3}$ and hence

$$E(X_{1,3}X_{3,3} \mid X_{2,3} = x) = E(X_{1,3} \mid X_{2,3} = x)E(X_{3,3} \mid X_{2,3} = x).$$

Chapter 19

Order Statistics and Record Values Based on F^α Distributions

In this chapter we study some distributional properties of order statistics and record values based on F^α distributions. The F^α distributions have been widely studied in statistics since 1995 because of their wide applicability in the modeling and analysis of life time data. Many researchers and authors have developed various classes of F^α distributions. Below we first briefly describe the definitions and properties of F^α distributions and review some selected members of F^α family of distributions. Then the distributional properties of record values and order statistics from these distributions are presented. For details, the interested readers are referred to Shakil and Ahsanullah (2012), and references therein.

19.1 F^α distributions

Definition 19.1. An absolutely continuous positive random variable X is said to have F^α distribution if its cdf is given by $G(x) = F^\alpha(x) = [F(x)]^\alpha$, $\alpha > 0$, which is the α-th power of the base line distribution function $F(x)$. The distribution $G(x)$ is also called an *exponentiated distribution of the given c.d.f.* $F(x)$. Its pdf $g(x)$ is given by $g(x) = \alpha f(x) F^{\alpha-1}(x)$, $\alpha > 0$, where $f(x) = \frac{dF(x)}{dx}$ denotes the pdf of the random variable X.

Remark 19.1. In literature, the $F^\alpha(x)$ distribution is also defined as the proportional reversed hazard rate model (PRHRM) due to the fact that the reversed hazard rate function (RHRF) of $G(x)$ is given by

$$\lambda_G^*(x) = \frac{d}{dx}\left(\ln(G(x))\right) = \frac{g(x)}{G(x)},$$

where $g(x)$ is the pdf corresponding to $G(x)$. Thus

$$\lambda_G^*(x) = \frac{\alpha(F(x))^{\alpha-1}f(x)}{(F(x))^\alpha} = \alpha\lambda_F^*(x),$$

M. Ahsanullah et al., *An Introduction to Order Statistics*,
Atlantis Studies in Probability and Statistics 3,
DOI: 10.2991/978-94-91216-83-1_19, © Atlantis Press 2013

which implies that the RHRF of $G(x)$ is proportional to the RHRF of $F(x)$ with real constant of proportionality α. When α is a positive integer, F^α is also defined as the Lehmann alternatives, that is, the model of non-parametric class of alternatives, see Lehmann (1953). The k-th moment of the random variable X having F^α distribution is given by

$$E[X^k] = \int_0^\infty x^k g(x)dx = \int_0^\infty x^k \left[\alpha f(x) F^{\alpha-1}(x)\right] dx. \tag{19.1}$$

Letting $F(x) = u$ in Eq. (19.1) and simplifying it, the expression for the k-th moment of F^α distribution is given by

$$E[X^k] = \alpha \int_0^1 \left[F^{-1}(u)\right]^k u^{\alpha-1} du, \tag{19.2}$$

where $F^{-1}(u)$ represents the inverse of the base distribution cdf $F(u)$ associated with F^α distribution. Thus, from (19.1), it is possible to determine the k-th moment of various distributions belonging to the family of F^α distributions provided the integral on the right side of the Eq. (19.2) exists and can be evaluated analytically in closed form or approximately.

19.2   Review of Some Members of the F^α Family of distributions

For the sake of completeness, here we briefly review some selected members of F^α family of distributions and their properties. See, for example, Shakil and Ahsanullah (2012), and references therein.

Gompertz-Verhulst Exponentiated distribution.   One of the earliest examples of F^α distribution found in literarture is Gompertz-Verhulst exponentiated distribution which is defined as follows:

$$G(x) = \left(1 - \rho \exp(-\lambda x)\right)^\alpha, \tag{19.3}$$

for $x > \dfrac{1}{\lambda} \ln \rho > 0$ and positive ρ, λ and α. The above model was used to compare known human mortality tables and to represent population growth. Some further generalization of Gompertz-Verhulst exponentiated distribution was considered by later researchers Note that, the exponentiated exponential distribution, also known as the generalized exponential distribution, as described below, is a particular case of Gompertz-Verhulst distribution (19.3) when $\rho = 1$.

Generalized Exponential (or Exponentiated Exponential (EE) distribution).   A random variable X is said to have the EE distribution if its cdf is given by

$$G(x) = (1 - \exp(-\lambda x))^\alpha, \tag{19.4}$$

for $x > 0$, $\lambda > 0$ and $\alpha > 0$, which is the α-th power of the cdf of the standard exponential distribution. The corresponding pdf of the EE distribution (19.4) is given by

$$g(x) = \alpha\lambda\exp(-\lambda x)\left[1 - \exp(-\lambda x)\right]^{\alpha-1}, \tag{19.5}$$

where α and λ are the shape and scale parameters, respectively. It is easy to see that the mean and variance of the random variable X with pdf (19.5) are, respectively, given by

$$E(X) = \Psi(\alpha+1) - \Psi(1),$$

and

$$V(X) = \Psi'(1) - \Psi'(\alpha+1),$$

where $\Psi(\cdot)$ and its derivative $\Psi'(\cdot)$ denote psi (or digamma) and polygamma functions, respectively.

Exponentiated Weibull (EW) distribution. A random variable X is said to have the exponentiated Weibull distribution if its cdf and pdf are respectively given by

$$G(x) = \left[1 - \exp(-\lambda x)^\delta\right]^\alpha, \tag{19.6}$$

and

$$\cdot g(x) = \alpha\delta\lambda^\delta x^{\delta-1}\left[1 - \exp(-\lambda x)^\delta\right]^{\alpha-1}, \tag{19.7}$$

for $x > 0 >$, $\lambda > 0$, $\delta > 0$ and $\alpha > 0$. The k-th moment of the EW random variable X with distribution function (19.6) is given by

$$E[X^k] = \alpha\lambda^{-k}\Gamma\left(\frac{k}{\delta}+1\right)\sum_{j=0}^{\infty}\frac{(1-\alpha)_j}{j!\,(j+1)^{(k-\delta)/\delta}},$$

for any k, where

$$(1-\alpha)_j = (-1)^j\frac{\Gamma(\alpha)}{\Gamma(\alpha-j)},$$

from which the mean and variance of the random variable X can easily be obtained.

Power Function distribution. A random variable X is said to have the power function distribution if its cdf and pdf are, respectively, given by

$$G(x) = \left(\frac{x}{\lambda}\right)^\alpha \tag{19.8}$$

and

$$g(x) = \frac{\alpha}{\lambda}\left(\frac{x}{\lambda}\right)^{\alpha-1}, \tag{19.9}$$

where $\lambda > 0$, $\alpha > 0$, and $0 < x < \lambda$.

Pareto Type II (or Exponentiated Pareto or Lomax) distribution. The Pareto type II distribution is defined by the cdf

$$G(x) = \left[1 - \frac{1}{1 + \lambda x^{\delta}}\right]^{\alpha}, \tag{19.10}$$

for $x > 0$, $\delta > 0$, $\lambda > 0$, and $\alpha > 0$. Another variant of the Pareto type II or the exponentiated Pareto distribution found in literature and introduced as a lifetime model has the cdf and pdf expressed, respectively, as

$$G(x) = \left[1 - \left(\frac{1}{1+x}\right)^{\beta}\right]^{\alpha} \tag{19.11a}$$

and

$$g(x) = \alpha\beta \left[1 - \left(\frac{1}{1+x}\right)^{\beta}\right]^{\alpha-1} \left(\frac{1}{1+x}\right)^{\beta+1}, \tag{19.11b}$$

for $x > 0$, $\beta > 0$, and $\alpha > 0$, where α and β denote the shape parameters of the exponentiated Pareto distribution given by (19.11a)-(19.11b). When $\alpha = 1$, the above distribution (19.10) corresponds to the standard Pareto distribution.

Burr Type X or Generalized Rayleigh (GR) distribution. The two-parameter Burr type X (also called the generalized Rayleigh distribution), denoted by $GR(\alpha, \lambda)$, where α and λ denote the shape and scale parameters respectively, used to model strength and lifetime data, has the cdf and pdf, respectively, expressed as

$$G(x) = \left(1 - \exp(-(\lambda x)^2)\right)^{\alpha} \tag{19.12}$$

and

$$g(x) = 2\alpha\lambda^2 x \exp(-\lambda x)^2 \left[1 - \exp(-(\lambda x)^2)\right]^{\alpha-1}, \tag{19.13}$$

for $x > 0$, $\lambda > 0$, and $\alpha > 0$. Note that the $GR(\alpha, \lambda)$ distribution is a special case of the EW distribution, as defined above. It is observed that for $\alpha \leqslant 0.5$, the pdf (19.13) of $GR(\alpha, \lambda)$ distribution is a decreasing function. It is right-skewed and unimodal for $\alpha > 0.5$. Further, the hazard function of $GR(\alpha, \lambda)$ distribution is bathtub type for $\alpha \leqslant 0.5$ and is increasing for $\alpha > 0.5$.

19.3 Other Members of F^{α} Family of distributions

It is evident from the above examples of F^{α} distributions that adding a parameter, say, $\alpha > 0$, to a cumulative distribution function (cdf) F by exponentiation produces a cdf F^{α} which is richer and more flexible to modeling data. It has been observed that F^α is flexible enough to accommodate both monotone as well as non-monotone hazard rates. For example, if F is exponential such that $F(x) = 1 - \exp(-\lambda x)$, then its pdf $f(x) = \lambda \exp(-\lambda x)$ is monotone decreasing on the positive half of the real line. However, $G(x) = (1 - \exp(-\lambda x))^\alpha$ has pdf $g(x) = \alpha \lambda \exp(-\lambda x)[1 - \exp(-\lambda x)]^{\alpha-1}$ which is unimodal on $(0, \infty)$ with mode at $x = (\ln)/\lambda$. Furthermore, while the exponential distribution F has constant hazard rate λ, it can be shown that the exponentiated exponential (EE) $G(x)$ has increasing hazard rate (IHD) if $\alpha > 1$, constant hazard rate if $\alpha = 1$, and decreasing hazard rate (DHR) if $\alpha < 1$. It follows from this analysis that adding one or more parameters to a distribution makes it richer and more flexible for modeling data. Obviously F^α distributions are quite different from F and need special investigation. It appears from literature that, since 1995, the F^α or exponentiated distributions have been widely studied by many authors. Besides the above stated F^α distributions, several other members of F^α family of distributions have recently been developed by various authors. For the sake of completeness, we briefly describe some of these members of the F^α family of distributions.

Generalized Logistic (GL) distribution. The two-parameter generalized logistic distribution, denoted by $GL(\alpha, \lambda)$, where α and λ denote the shape and scale parameters, respectively, has the cdf and pdf, respectively, expressed as

$$G(x) = \frac{1}{(1 + e^{-\lambda x})^\alpha} \tag{19.14}$$

and

$$g(x) = \frac{\alpha \lambda e^{-\lambda x}}{(1 + e^{-\lambda x})^{\alpha+1}}, \tag{19.15}$$

for $-\infty < x < \infty$, $\lambda > 0$, and $\alpha > 0$. The $GL(\alpha, \lambda)$ distribution was originally proposed as a generalization of the logistic distribution. It is observed that $GL(\alpha, \lambda)$ is skewed and its kurtosis coefficient is greater than that of the logistic distribution. For $\alpha = 1$, the GL distribution becomes the standard logistic and is symmetric. Further, it is observed that the pdf (19.15) of $GL(\alpha, \lambda)$ distribution is increasing for $x < \lambda^{-1} \ln \alpha$ and is decreasing for $x > \lambda^{-1} \ln \alpha$. Therefore it is a unimodal distribution with mode at $\ln \alpha$. The density function of $GL(\alpha, \lambda)$ is log-concave for all values of α. For $\alpha > 1$, it is positively skewed, and for $0 < \alpha < 1$, it is negatively skewed. For this, the proportionality constant α can be represented as the skewness parameter. The hazard rate of GL distribution can be either bathtub type or an increasing function, depending on the shape parameter α. For $\alpha < 1$, the hazard rate of $GL(\alpha, \lambda)$ is bathtub type and for $\alpha \geqslant 1$, it is an increasing function.

For other members of members of F^α family of distributions recently introduced in literature by researchers, such as the exponentiated generalized inverse Gaussian (EGIG) distribution, the exponentiated gamma (EGamma) distribution, exponentiated Fréchet distribution, exponentiated Gumbel distribution, exponentiated extreme value distribution, exponentiated exponential geometric distribution (also known as the generalized exponential geometric (GEG) distribution), exponentiated inverse Weibull distribution, the interested readers are referred to the references as provided in Shakil and Ahsanullah (2012).

19.4   Record Values and Order Statistics from F^α distributions

For our ready reference, we will first recall the distributions of order statistics and record values as given below.

Order Statistics.   Suppose that (X_r) $(r = 1, 2, \ldots, n)$ is a sequence of n independent and identically distributed (*i.i.d.*) random variables (*rv's*), each with cdf $F(x)$. If these are rearranged in ascending order of magnitude and written as $X_{1,n} \leqslant X_{2,n} \leqslant \cdots \leqslant X_{n,n}$, then $X_{r,n}$ $(r = 1, 2, \ldots, n)$, is called the r-th order statistic from a sample of size n. Further $X_{1,n}$ and $X_{n,n}$ are called extreme order statistics, and $R = X_{n,n} - X_{1,n}$ is called the range. If $F_{r,n}(x)$ and $f_{r,n}(x)$ $(r = 1, 2, \ldots, n)$, denote the cdf and pdf of the r-th order statistic $X_{r,n}$ respectively, then these are defined as

$$F_{r,n}(x) = \sum_{j=r}^{n} \binom{n}{j} [F(x)]^j [1 - F(x)]^{n-j} = I_{F(x)}(r, n - r + 1), \qquad (19.16)$$

and

$$f_{r,n}(x) = \frac{1}{B(r, n - r + 1)} [F(x)]^{r-1} [1 - F(x)]^{n-r} f(x) \qquad (19.17)$$

where

$$I_p(a, b) = \frac{1}{B(a, b)} \int_0^p t^{a-1} (1 - t)^{b-1} dt$$

denotes the incomplete Beta function.

Record Values.   Suppose that $(X_n)_{n \geqslant 1}$ is a sequence of *i.i.d.* *rv's* with cdf $F(x)$. Let $Y_n = \max(\min) \{X_j \mid 1 \leqslant j \leqslant n\}$ for $n \geqslant 1$. We say X_j is an upper (lower) record value of $\{X_n \mid n \leqslant 1\}$ if $Y_j > (<)Y_{j-1}$, $j > 1$. By definition X_1 is an upper as well as a lower record value.

Lower Record Values. The indices at which the lower record values occur are given by the record times $\{\ell(n),\ n \geqslant 1\}$, where $\ell(n) = \min\{j \mid j > \ell(n-1),\ X_j < X_{\ell(n-1)},\ n \leqslant 1\}$ and $\ell(1) = 1$. We will denote $\ell(n)$ as the indices where the lower record values occur. The n-th lower record value will be denoted by $X_{\ell(n)}$. If we define $F_{(n)}$ as the c.d.f. of $X_{\ell(n)}$ for $n \geqslant 1$, then we have

$$F_{(n)} = \int_{-\infty}^{x} \frac{(H(u))^{n-1}}{\Gamma(n)} dF(u), \quad -\infty < x < \infty, \tag{19.18}$$

where $H(x) = -\ln F(x)$ and $h(x) = -\frac{d}{dx} H(x) = f(x)(F(x))^{-1}$. The pdf of $X_{L(n)}$, denoted by $f_{(n)}$, is

$$f_{(n)}(x) = \frac{(H(x))^{n-1}}{\Gamma(n)} f(x), \quad -\infty < x < \infty, \tag{19.19}$$

Upper Record Values. The indices at which the upper record values occur are given by the record times $\{L(n),\ n \geqslant 1\}$, where $L(n) = \min\{j \mid j > L(n-1),\ X_j > X_{L(n-1)},\ n > 1\}$ and $L(1) = 1$. Many properties of the upper record value sequence can be expressed in terms of the cumulative hazard rate function $R(x) = -\ln \overline{F}(x)$, where $\overline{F}(x) = 1 - F(x)$, $0 < \overline{F}(x) < 1$. If we define $F_n(x)$ as the cdf of $X_{L(n)}$ for $n \geqslant 1$ then we have

$$F_n(x) = \int_{-\infty}^{x} \frac{(R(u))^{n-1}}{\Gamma(n)} dF(u), \quad -\infty < x < \infty, \tag{19.20}$$

from which it is easy to see that

$$F_n(x) = 1 - \overline{F}(x) \sum_{j=0}^{n-1} \frac{(R(x))^j}{\Gamma(j+1)},$$

that is,

$$\overline{F}_n(x) = \overline{F}(x) \sum_{j=0}^{n-1} \frac{(R(x))^j}{\Gamma(j+1)}.$$

The pdf of $X_{U(n)}$, denoted by f_n is

$$f_n(x) = \frac{(R(x))^{n-1}}{\Gamma(n)} f(x), \quad -\infty < x < \infty. \tag{19.21}$$

It is easy to see that $\overline{F}_n(x) - \overline{F}_{n-1}(x) = \overline{F}(x) \frac{f_n(x)}{f(x)}$. Using Eq. (19.19), the pdf $f_{(n)}(x)$ cdf $F_{(n)}(x)$ of the n-th lower record value $X(n)$ from F^α distribution are, respectively, given by

$$f_{(n)}(x) = \frac{[-\alpha \ln F(x)]^{n-1} \alpha (F(x))^{\alpha-1} f(x)}{\Gamma(n)}, \quad n = 1, 2, \ldots, \tag{19.22}$$

and

$$F_{(n)}(x) = \frac{\Gamma(n, -\alpha \ln F(x))}{\Gamma(n)}, \quad n = 1, 2, \ldots, \tag{19.23}$$

where $\alpha > 0$, and $\Gamma(c,z) = \int_z^\infty e^{-t}t^{c-1}dt$, $c > 0$, denotes incomplete gamma function. The k-th moment of the n-th lower record value $X(n)$ with the pdf (19.22) is given by

$$E[X^k(n)] = \int_0^\infty x^k \frac{[-\ln(F^\alpha(x))]^{n-1}[\alpha f(x)F^{\alpha-1}(x)]}{\Gamma(n)}\,dx. \qquad (19.24)$$

Letting $-\ln(F^\alpha(x)) = u$ in Eq. (19.24) and simplifying it, the expression for the k-th moment of the n-th lower record value $X(n)$ is easily obtained as

$$E[X^k(n)] = \frac{1}{\Gamma(n)}\int_0^\infty \left[F^{-1}\left(\exp\left(-\frac{u}{\alpha}\right)\right)\right]^k u^{n-1}\exp(-u)du, \qquad (19.25)$$

where $F^{-1}(z)$ represents the inverse of the base distribution cdf $F(z)$ associated with F^α distribution. Thus, from (19.25), it is possible to determine the k-th moment of the n-th lower record value $X(n)$ from various distributions belonging to the family of F^α distribution provided the integral on the right side of the Eq. (19.25) exists and can be evaluated analytically in closed form or approximately.

Using Eq. (19.16) and Eq. (19.17), the pdf $f_{r,n}(x)$ and the cdf $F_{r,n}(x)$ of the r-th order statistics and r moment of the r-th order statistic $X_{r,n}$, $1 \leqslant r \leqslant n$, from F^α distribution, can similarly be expressed as

$$F_{r,n}(x) = \sum_{j=r}^n \binom{n}{j}[F(x)]^{\alpha r-1}[1 - F^\alpha(x)]^{n-r},$$

$$f_{r,n}(x) = \frac{\alpha}{B(r,n-r+1)}[F(x)]^{\alpha r-1}[1 - F^\alpha(x)]^{n-r},$$

and

$$E[X^k(n)] = \frac{1}{B(r,n-r+1)}\int_0^1 u^{r-1}(1-u)^{n-r}[F^{-1}(u^{1/\alpha})]^k du,$$

where $F^\alpha(x) = u$.

In what follows, we will briefly discuss some distributional properties of record values and order statistics from various distributions belonging to F^α family of distributions. For details, the interested readers are referred to Shakil and Ahsanullah (2012), and references therein.

19.4.1 Generalized Exponential (GE) or Exponentiated Exponential (EE) distribution

Order Statistics: Let $X_1, X_2, \ldots, X_n$ be a random sample from the GE (EE) distribution with cdf and pdf as (19.4) and (19.5) respectively, and let $X_{1,n} \leqslant X_{2,n} \leqslant \cdots \leqslant X_{n,n}$ denote the order statistics obtained from this sample. One can easily obtain the pdf $f_{r,n}(x)$ and cdf $F_{r,n}(x)$ of the r-th order statistic $X_{r,n}$, $1 \leqslant r \leqslant n$, from the GE (EE) distribution by using

Eqs. (19.4) and (19.5) in (19.16) and (19.17) respectively. When $\lambda = 1$, the pdf of the r-th order statistic $X_{r,n}$, $1 \leqslant r \leqslant n$, for $x > 0$, is easily obtained as

$$f_{r,n}(x) = \sum_{j=0}^{n-r} d_j(n,r) f(x; \alpha_{r+j}),$$

where

$$\alpha_j = j\alpha, \quad d_j(n,r) = \frac{(-1)^j n \binom{n-1}{r-1}\binom{n-r}{j}}{r+j},$$

and $f(x, \alpha_{r+j})$ denotes the EE distribution with the shape parameter $\alpha(r+j)$ and scale parameter 1. Note that the coefficients $d_j(n,r)$ $(j = 1, 2, \ldots, n-r)$, are not dependent on α. The moment generating function, $M_{X_{r,n}}(t)$, of $X_{r,n}$, for $|t| < 1$, is easily given by

$$M_{X_{r,n}}(t) = E(\exp(tX_{r,n})) = \int_0^\infty e^{tx} f_{r,n}(x) dx$$

$$= \frac{\alpha}{B(r, n-r+1)} \sum_{j=0}^{n-r} (-1)^j \binom{n-r}{j} \frac{\Gamma(\alpha(r+j))\Gamma(1-t)}{\Gamma(\alpha(r+j)-t+1)}, \quad |t| < 1.$$

Differentiating $M_{X_{r,n}}(t)$ twice with respect to t and evaluating at t=0, it is easy to see that the mean (first moment) and the second moment of $X_{r,n}$ $(r = 1, 2, \ldots, n)$, are, respectively, given by

$$\alpha_{r,n} = E(X_{r,n}) = \frac{1}{B(r, n-r+1)} \sum_{j=0}^{n-r} (-1)^j \frac{\binom{n-r}{j}}{r+j} [\Psi(\alpha(r+j)+1) + \gamma],$$

and

$$\alpha_{r,n}^{(2)} = E(X_{r,n}^2)$$

$$= \frac{1}{B(r, n-r+1)} \sum_{j=0}^{n-r} (-1)^j \frac{\binom{n-r}{j}}{r+j} [\{\Psi(\alpha(r+j)+1)\}^2 + \Psi'(1) - \Psi'(\alpha(r+j)+1]$$

where $\psi(z) = \frac{d}{dz} \ln \Gamma(z)$ denotes digamma function, $\gamma = \Psi(1) = 0.577215$ is the Euler's constant, and $\Psi'(1) = \xi(2) = \pi^2/6$, where $\Psi(z)$ and $\xi(z)$ are called trigamma and Riemann zeta functions, respectively.

Record Values. The pdf f_n and cdf F_n of the n-th lower record value $X(n)$ from EE distribution can easily be obtained by using Eqs. (19.4) and (19.5) in (19.18) and (19.9) respectively. When $\lambda = 1$, the moment generating function, $M(t)$, of the n-th lower record value $X(n)$ is easily given by

$$M_{X(n)}(t) = E(\exp(t)X(n)) = \int_0^\infty \exp(tx) f_n dx = \alpha^n \sum_{j=0}^\infty \frac{(t)_j}{j!} \frac{1}{(\alpha+j)^n}, \quad t \neq 0,$$

where $(t)_j = t(t+1)\cdots(t+j-1)$, $j = 1, 2, 3, \ldots$, and $(t)_0 = 1$.

Thus differentiating $M_{X(n)}(t)$ and evaluating at $t = 0$, the mean and the second moment of the n-th lower record value $X(n)$ are easily obtained as

$$E(X(n)) = \sum_{j=1}^{\infty} \frac{\alpha^n}{j(\alpha+j)^n}, \quad n \geqslant 1,$$

and

$$E(X^2(n)) = \alpha^n \sum_{j=1}^{\infty} \sum_{k=1}^{\infty} \frac{\alpha^n}{jk(j+k+\alpha)}.$$

19.4.2 *Exponentiated Weibull (EW) distribution*

The pdf f_n and cdf F_n of the n-th lower record value $X(n)$ from EW distribution, by using Eqs. (19.6) and (19.7) in (19.18) and (19.9) respectively, are easily given as

$$f_n(x) = \frac{\left[-\ln([1-\exp(-(\lambda x)^\delta)]^\alpha)\right]^{n-1}\left[\alpha\delta\lambda^\delta x^{\delta-1}\exp(-\lambda x)^\delta[1-\exp(-(\lambda x)^\delta)]\right]^{\alpha-1}}{\Gamma(n)},$$

and

$$F_n(x) = \frac{\Gamma\left(n, -\ln([1-\exp(-(\lambda x))^\delta]^\alpha)\right)}{\Gamma(n)},$$

where $n > 0$ (an integer), $x > 0$, $\lambda > 0$, $\delta > 0$, $\alpha > 0$, and $\Gamma(c,z) = \int_z^\infty e^{-t}t^{c-1}dt$, $c > 0$, denotes incomplete gamma function. The k-th moment of the n-th lower record value $X(n)$ from EW distribution is given by

$$E[X^k(n)] = \int_0^\infty x^k \frac{\left[-\ln([1-\exp(-(\lambda x)^\delta)]^\alpha)\right]^{n-1}}{\Gamma(n)}$$
$$\times \left[\alpha\delta\lambda^\delta x^{\delta-1}\exp(-\lambda x)^\delta[1-\exp(-(\lambda x)^\delta)]\right]^{\alpha-1} dx.$$

Letting $-\ln([1-\exp(-(\lambda x)^\delta)]^\alpha) = u$ in the above equation and simplifying it, the expression for the k-th moment of the n-th lower record value $X(n)$ is obtained as

$$E[X^k(n)] = \frac{1}{\lambda^k\Gamma(n)} \int_0^\infty \left[-\ln\left(1-\exp\left(-\frac{u}{\alpha}\right)\right)\right]^{k/\delta} u^{n-1}\exp(-u)du,$$

which, in view of the fact that

$$-\ln\left(1-\exp\left(-\frac{u}{\alpha}\right)\right) = \sum_{j=1}^{\infty} \frac{\exp(-ju/\alpha)}{j},$$

reduces to

$$E[X^k(n)] = \frac{1}{\lambda^k\Gamma(n)} \int_0^\infty \left[\sum_{j=1}^{\infty} \frac{\exp(-ju/\alpha)}{j}\right]^{k/\delta} u^{n-1}\exp(-u)du,$$

It is evident from the above expression that the analytical derivations of the exact or tractable closed form expressions for the moments of record statistics from EW distribution seem to be complicated, and therefore need further investigations. However, for $\delta = 1$, since EW distribution reduces to EE distribution, one can easily derive the moments of record statistics in exact form. Similarly, one can obtain the pdf $f_{(r,n)}(x)$ and cdf $F_{(r,n)}(x)$ of the r-th order statistic $X_{(r,n)}$, $1 \leqslant r \leqslant n$, from EW distribution by using Eqs. (19.6) and (19.7) in (19.16) and (19.17) respectively. Note that the analytical derivations of the exact or tractable closed form expressions for the moments of order statistics from EW distribution are complicated, and therefore also need further investigations.

19.4.3 *Power-Function distribution*

The pdf f_n and cdf F_n of the n-th lower record value $X(n)$ from power function distribution can easily obtained by using Eqs. (19.8) and (19.9) in (19.18) and (19.19) respectively, as

$$f_n(x) = \frac{[-\ln(x/\lambda)^\alpha]^{n-1}[(\alpha/\lambda)(x/\lambda)^{\alpha-1}]}{\Gamma(n)}$$

and

$$F_n(x) = \frac{\Gamma(n, -\ln(x/\lambda)^\alpha)}{\Gamma(n)},$$

where $n > 0$ (an integer), $0 < x < \lambda$, $\lambda > 0$, and $\alpha > 0$, and $\Gamma(c,z) = \int_z^\infty e^{-t} t^{c-1} dt$, $c > 0$, denotes incomplete gamma function. The k-th moment of the n-th lower record value $X(n)$ from power function distribution is easily given by

$$E[X^k(n)] = \int_0^\lambda x^k \frac{[-\ln(x/\lambda)^\alpha]^{n-1}[(\alpha/\lambda)(x/\lambda)^{\alpha-1}]}{\Gamma(n)} dx.$$

Letting $-\ln(x/\lambda)^\alpha = u$ in the above equation and simplifying it, the expression for the k-th moment of the n-th lower record value $X(n)$ is easily obtained as

$$E[X^k(n)] = \frac{\lambda^k \alpha^n}{(\alpha+k)^n},$$

from which the single moment (that is, the mean), the second single moment and variance of the n-th lower record value $X(n)$ from power function distribution are easily given as

$$E[X(n)] = \frac{\lambda \alpha^n}{(\alpha+1)^n},$$

$$E[X^2(n)] = \frac{\lambda^2 \alpha^n}{(\alpha+2)^n},$$

and

$$\text{Var}[X(n)] = \lambda^2 \alpha^n \left[\frac{1}{(\alpha+2)^n} - \frac{\alpha^n}{(\alpha+1)^{2n}} \right].$$

Similarly, one can obtain the pdf $f_{(r,n)}(x)$ and cdf $F_{(r,n)}(x)$ of the r-th order statistic $X_{(r,n)}$, $1 \leqslant r \leqslant n$, from power function distribution by using Eqs. (19.8) and (19.9) in (19.16) and (19.17) respectively. We can also derive expressions for the moments of order statistics from power function distribution (19.8). Thus, if the random variable X has the pdf (19.9), the k-th moment of the n-th order statistic $X_{(r,n)}$ from a sample of size n is given by

$$E\left(X_{(r,n)}^k\right) = \frac{\Gamma(n+1)\,\Gamma\left(\frac{k}{\alpha}+r\right)\lambda^k}{\Gamma(r)\,\Gamma\left(n+\frac{k}{\alpha}+1\right)}, \quad r = 1, 2, 3, \ldots, n,$$

where $\Gamma(z)$ denotes gamma function. Also we have the following recurrence relation between moments of the order statistics $X_{(r,n)}$ and $X_{(r-1,n)}$:

$$E\left(X_{(r,n)}^k\right) = \frac{\left(r+\frac{k}{\alpha}-1\right)E\left(X_{(r-1,n)}^k\right)}{(r-1)}, \quad r = 1, 2, 3, \ldots, n, \ k > 1.$$

Using the above equations, the single moment (that is, the mean), the second single moment and variance of the r-th order statistic, $X_{r,n}$, $1 \leqslant r \leqslant n$, from power function distribution, can easily be derived.

19.4.4 *Pareto Type II (or Exponentiated Pareto or Lomax) distribution*

The pdf f_n and cdf F_n of the n-th lower record value $X(n)$ from $EP(\alpha, \beta)$ can easily be obtained by using the Eqs. (19.11a) and (19.11b) in (19.18) and (19.19) respectively. Similarly, one can obtain the pdf $f_{r,n}(x)$ and cdf $F_{r,n}(x)$ of the r-th order statistic $X_{r,n}$, $1 \leqslant r \leqslant n$, from $EP(\alpha, \beta)$ by using the Eqs. (19.11a) and (19.11b) in (19.16) and (19.17) respectively. It appears from the literature that not much attention has been paid on the order statistics and record values from $EP(\alpha, \beta)$, and therefore need further investigations.

19.4.5 *Burr Type X or Generalized Rayleigh (GR) distribution*

By using Eqs. (19.12) and (19.13) in (19.18) and (19.19) respectively, the pdf f_n and cdf F_n of the n-th lower record value $X(n)$ from $GR(\alpha, \lambda)$ distribution can easily be obtained as

$$f_n(x) = \frac{\left[-\ln\left(\left[1-\exp(-(\lambda x)^2)\right]^\alpha\right)\right]^{n-1}\left[2\alpha\lambda^2 x \exp(-\lambda x)^2\left[1-\exp(-(\lambda x)^2)\right]^{\alpha-1}\right]}{\Gamma(n)},$$

and

$$F_n(x) = \frac{\Gamma\left(n, -\ln\left(\left[1-\exp(-(\lambda x)^2)\right]^\alpha\right)\right)}{\Gamma(n)},$$

where $n > 0$ (an integer), $x > 0$, $\lambda > 0$, $\alpha > 0$, and $\Gamma(c,z) = \int_z^\infty e^{-t}t^{c-1}dt, c > 0$, denotes incomplete gamma function. The k-th moment of the n-th lower record value $X(n)$ from $GR(\alpha,\lambda)$ distribution is given by

$$E[X^k(n)] = \int_0^\infty x^k \frac{1}{\Gamma(n)}dx.$$

Letting $-\ln([1 - \exp(-(\lambda x))]^\alpha) = u$ in the above equation and simplifying it, the expression for the k-th moment of the n-th lower record value $X(n)$ is obtained as

$$E[X^k(n)] = \frac{1}{\lambda^k \Gamma(n)} \int_0^\infty \left[-\ln\left(1 - \exp\left(-\frac{u}{\alpha}\right)\right)\right]^{k/2} u^{n-1} \exp(-u)du,$$

which, in view of the fact that

$$-\ln\left(1 - \exp\left(-\frac{u}{\alpha}\right)\right) = \sum_{j=1}^\infty \frac{\exp(-ju/\alpha)}{j},$$

reduces to

$$E[X^k(n)] = \frac{1}{\lambda^k \Gamma(n)} \int_0^\infty \left[\sum_{j=1}^\infty \frac{\exp(-ju/\alpha)}{j}\right]^{k/2} u^{n-1} \exp(-u)du,$$

It is evident from the above expression that the analytical derivation of the exact or tractable closed form expressions for the moments of record statistics from $GR(\alpha,\lambda)$ distribution is complicated, and, therefore, needs further investigations. By using Eqs. (19.12) and (19.13) in (19.16) and (19.17) the pdf $F_{r,n}(x)$ of the r-th order statistic $X_{r,n}$, $1 \leqslant r \leqslant n$, from $GR(\alpha,\lambda)$ distribution are given by

$$f_{r,n}(x) = \frac{1}{B(r,n-r+1)} \left[\frac{\Gamma(n, -\ln([1 - \exp(-(\lambda x)^2)]^\alpha))}{\Gamma(n)}\right]^{r-1}$$

$$\times \left[1 - \frac{\Gamma(n, -\ln([1 - \exp(-(\lambda x)^2)]^\alpha))}{\Gamma(n)}\right]^{n-r}$$

$$\times \frac{[-\ln([1 - \exp(-(\lambda x)^2)]^\alpha)]^{n-1}[2\alpha\lambda^2 x \exp(-\lambda x)^2 [1 - \exp(-(\lambda x)^2)]^{\alpha-1}]}{\Gamma(n)}.$$

It appears from the literature that not much attention has also been paid on the record values from generalized Rayleigh distribution, and therefore need further investigations.

19.4.6 Record Values and Order Statistics from Other Members of F^α Family of distributions

Besides the above stated F^α distributions, though several other members of F^α family of distributions have recently been developed by various authors, such as the exponentiated generalized inverse Gaussian (EGIG) distribution, the exponentiated gamma (EGamma) distribution, exponentiated Fréchet distribution, exponentiated Gumbel distribution, exponentiated extreme value distribution, exponentiated exponential geometric distribution (also known as the generalized exponential geometric (GEG) distribution), exponentiated inverse Weibull distribution, see, for example, Shakil and Ahsanullah (2012) and references therein, it appears from literature that not much attention has been paid to the analysis of order statistics and record values from these distributions, and, therefore, needs further and special investigations.

19.5   Concluding Remarks and Future Directions of Research

It is evident from the above examples of F^α distributions that adding a parameter, say, $\alpha > 0$, to a cdf F by exponentiation produces a cdf F^α which is richer and more flexible to modeling data. It has been observed that F^α is flexible enough to accommodate both monotone as well as non-monotone hazard rates. For example, if F is exponential such that $F(x) = 1 - \exp(-\lambda x)$, then its pdf $f(x) = \lambda \exp(-\lambda x)$ is monotone decreasing on the positive half of the real line. However, $G(x) = (1 - \exp(-\lambda x))^\alpha$ has pdf $g(x) = \alpha \lambda \exp(-\lambda x)[1 - \exp(-\lambda x)]^{\alpha-1}$ which is unimodal on $(0, \infty)$ with mode at $x = (\ln \alpha)/\lambda$. Furthermore, while the exponential distribution F has constant hazard rate λ, it can be shown that the exponentiated exponential (EE) G has increasing hazard rate (IHD) if $\alpha > 1$, constant hazard rate if $\alpha = 1$ and decreasing hazard rate (DHR) if $\alpha < 1$. It follows from this analysis that adding one or more parameters to a distribution makes it richer and more flexible for modeling data. Obviously F^α distributions are quite different from F and need special investigations. It appears from literature that, since 1995, many researchers and authors have studied and developed various classes of F^α distributions, among them, the exponentiated exponential and exponentiated Weibull distributions are notable. However, in spite of the extensive recent work on F^α distributions, characterizations, estimation of parameters and their applications, very little attention has been paid to the study of order statistics and record values, prediction of future order statistics and record values based on existing ones, and inferential properties of order statistics and record values from F^α family of distributions. Therefore, the purpose of this chapter was to study some of the distributional properties of order statistics and record values from F^α or exponentiated family of distributions. Most of the recent works are reviewed and discussed. The expressions for the pdf, cdf and moments for certain members of F^α family of distributions and associated order statistics and record values are provided. We hope that the findings of this material will be useful for the practitioners in various fields of studies and further enhancement of research in order statistics, record value theory and their applications. These results will also be useful in the study of reversed hazard rates. One can also study the applications of order statistics and record values in the statistical data analysis when the parent distributions belong to the F^α family of distributions. It will be useful in the characterizations of F^α distributions and quantifying information contained in observing each order statistic and record value. For future work, one can consider to develop inferential procedures for the parameters of the parent F^α distributions based on the corresponding order statistics and record value distributions. It is also of theoretical interest, in designing an experiment, to investigate and establish the relationships between the record values and the order statistics when observations are taken from the same parent F^α distributions

Exercise 19.1. Discuss the distributional properties of $X_{i,n}$ $(1 \leqslant i \leqslant n)$, the i-th order statistics based on a random sample of size n from a generalized exponential distribution.

Exercise 19.2. Discuss the distributional properties of the n-th record value based on a random sample of size n from a generalized exponential distribution.

Exercise 19.3. Discuss the distributional properties of $X_{i,n}$ $(1 \leqslant i \leqslant n)$, the i-th order statistics based on a random sample of size n from a generalized Weibull distribution.

Exercise 19.4. Discuss the distributional properties of the n-th record value based on a random sample of size n from a generalized Weibull distribution.

Exercise 19.5. Discuss the distributional properties $X_{i,n}$ $(1 \leqslant i \leqslant n)$, the i-th order statistics based on a random sample of size n when the underlying distributions are following:

 (i) a power function distribution;
 (ii) an exponentiated Pareto (or Lomax) distribution;
(iii) a generalized Rayleigh distribution;
(iv) a generalized logistic distribution;
 (v) an exponentiated generalized inverse Gaussian (EGIG) distribution.

Exercise 19.6. Discuss the distributional properties of the n-th record value based on a random sample of size n when the underlying distributions are following:

 (i) a power function distribution;
 (ii) an exponentiated Pareto (or Lomax) distribution;

(iii) a generalized Rayleigh distribution;

(iv) a generalized logistic distribution;

 (v) an exponentiated generalized inverse Gaussian (EGIG) distribution.

Exercise 19.7. Estimate the parameters of the order statistics from $EP(\alpha, \beta)$, and compared the performances of different estimation procedures through numerical simulations. *Hint*: See Shawky and Abu-Zinadah (2009).

Exercise 19.8. Discuss the distributional properties and estimation problems of the first order statistics from exponentiated Pareto distribution $EP(\alpha, \beta)$. *Hint*: See Ali and Woo (2010).

Exercise 19.9. Determine the exact forms of the pdf, moments of single, double, triple and quadruple of lower record values and recurrence relations between single, double, triple and quadruple moments of lower record values from from $EP(\alpha, \beta)$. *Hint*: See Shawky and Abu-Zinadah (2008).

Exercise 19.10. Determine the exact forms of the pdf, moments, measures of skewness and kurtosis for the order statistics from $GR(\alpha, 1)$ distribution. Estimate the parameters of order statistics from $GR(\alpha, 1)$ distribution. Also, tabulate the percentage points order statistics from $GR(\alpha, 1)$ distribution for proper choices of parameter. *Hint*: See Raqab (1998).

Exercise 19.11. Discuss the distributional properties of the n-th record value based on a random sample of size n when the underlying distribution is an exponentiated gamma distribution. *Hint*: See Shawky and Bakoban (2008).

Chapter 20

Generalized Order Statistics

Below we present the so-called generalized order statistics of several univariate absolutely continuous distributions.

Definition 20.1. Kamps (1995) introduced the generalized order statistics (gos). The order statistics, record values and sequential order statistics are special cases of this generalized order statistics. Suppose $X(1,n,m,k),\ldots,X(n,n,m,k)$ ($k \geqslant 1$, m is a real number), are n generalized order statistics. Then the joint pdf $f_{1,\ldots,n}(x_1,\ldots,x_n)$ can be written as

$$f_{1,\ldots,n}(x_1,\ldots,x_n) = \begin{cases} k\prod_{j=1}^{n-1}\gamma_j \prod_{i=1}^{n-1} (1-F(x_i))^m f(x_i)(1-F(x_n))^{k-1}f(x_n), \\ \qquad\qquad \text{for } F^{-1}(0) < x_1 < \cdots < x_n < F^{-1}(1), \\ 0, \qquad\quad \text{otherwise,} \end{cases} \tag{20.1}$$

where $\gamma_j = k + (n-j)(m+1)$ and $f(x) = \dfrac{dF(x)}{dx}$.

If $m = 0$ and $k = 1$, then $X(r,n,m,1)$ reduces to the ordinary r-th order statistic and (20.1) is the joint pdf of the n order statistics $X_{1,n} \leqslant \cdots \leqslant X_{n,n}$. If $k = 1$ and $m = -1$, then (20.1) is the joint pdf of the first n upper record values of the independent and identically distributed random variables with distribution function $F(x)$ and the corresponding probability density function $f(x)$.

Example 20.1. Let $F_{r,n,m,k}(x)$ be the distribution function of $X(r,n,m,k)$. Then the following equalities are equal:

$$F_{r,n,m,k}(x) = I_{\alpha(x)}\left(r, \frac{\gamma_r}{m+1}\right), \quad \text{if } m > -1, \tag{20.2}$$

and

$$F_{r,n,m,k}(x) = \Gamma_{\beta(x)}(r), \qquad \text{if } m = -1, \tag{20.3}$$

M. Ahsanullah et al., *An Introduction to Order Statistics*,
Atlantis Studies in Probability and Statistics 3,
DOI: 10.2991/978-94-91216-83-1_20, © Atlantis Press 2013

where

$$\alpha(x) = 1 - (\overline{F}(x))^{m+1}, \quad \overline{F}(x) = 1 - F(x),$$

$$\beta(x) = -k\ln\overline{F}(x),$$

$$I_x(p,q) = \frac{1}{B(p,q)} \int_0^x u^{p-1}(1-u)^{q-1}du, \quad B(p,q) = \frac{\Gamma(p)\Gamma(q)}{\Gamma(p+q)},$$

and

$$\Gamma_x(r) = \int_0^x \frac{1}{\Gamma(r)} u^{r-1}e^{-u}du.$$

Proof.   For $m > -1$, it follows from (20.1) that

$$F_{r,n,m,k}(x) = \int_{F^{-1}(0)}^x \frac{c_r}{(r-1)!}(1-F(u))^{k+(n-r)(m+1)-1}g_m^{r-1}(F(u))f(u)du,$$

where $c_r = \prod_{j=1}^r \gamma_j$.

Using the relation

$$B\left(r, \frac{\gamma_r}{m+1}\right) = \frac{\Gamma(r)(m+1)^r}{c_r}$$

and substituting $u = 1 - (\overline{F}(x))^{m+1}$, we get on simplification

$$F_{r,n,m,k}(x) = \frac{1}{B\left(r, \frac{\gamma_r}{m+1}\right)} \int_0^{1-(\overline{F}(x))^{m+1}} (1-u)^{\gamma_r-1}(1-u)^{r-1}du$$

$$= I_{\alpha(x)}\left(r, \frac{\gamma_r}{m+1}\right).$$

For $m = -1$ the following relations are valid:

$$F_{r,n,m,k}(x) = \int_{F^{-1}(0)}^x \frac{k^r}{(r-1)!}(1-F(u))^{k-1}(-\ln(1-F(u)))^{r-1}f(u)du$$

$$= \int_0^{-k\ln\overline{F}(x)} \frac{1}{(r-1)!}t^{r-1}e^{-t}dt$$

$$= \Gamma_{\beta(x)}(r), \quad \beta(x) = -k\ln\overline{F}(x). \qquad \square$$

Exercise 20.1. Prove that the following equality is valid:

$$F_{r,n,m,k}\left(r, \frac{\lambda_r}{m+1}\right) - F_{r,n,m,k}\left(r+1, \frac{\lambda_{r+1}}{m+1}\right) = \frac{1}{\gamma_{r+1}} \frac{\overline{F}(x)}{f(x)} f_{r+1,n,m,k}.$$

Example 20.2 (Gos of exponential distribution). Let X be a random variable (r.v.) whose probability density function (pdf) f is given by

$$f(x) = \begin{cases} \sigma^{-1}\exp(-\sigma^{-1}x), & \text{for } x > 0, \ \sigma > 0, \\ 0, & \text{otherwise.} \end{cases} \qquad (20.4)$$

We will denote $X \in E(\sigma)$ if the pdf of X is of the form given by (20.2).

If $X_i \in E(\sigma)$, $i = 1, 2, \ldots$, and X's are independent, then $X(1, n, m, k) \in E(\sigma/\gamma_1)$ and

$$X(r+1, n, m, k) \overset{d}{=} X(r, n, m, k) + \sigma \frac{W}{\gamma_{r+1}}, \quad r > 1,$$

where W is independent of $X(r, n, m, k)$ and $W \in E(1)$.

Proof. Integrating out $x_1, \ldots, x_{r-1}, x_{r+1}, \ldots, x_n$ from (20.1), we get the pdf $f_{r,n,m,k}$ of $X(r, n, m, k)$ as

$$f_{r,n,m,k}(x) = \frac{c_r}{(r-1)!} (1 - F(u))^{k+(n-r)(m+1)-1} g_m^{r-1}(F(u)) f(u), \qquad (20.5)$$

where

$$c_r = \prod_{j=1}^{r} \gamma_j$$

and

$$g_m(x) = \begin{cases} \dfrac{1}{m+1} (1 - (1-x)^{m+1}), & m \neq -1, \\ -\ln(1-x), & m = -1, x \in (0, 1). \end{cases}$$

Since

$$\lim_{m \to -1} \frac{1}{m+1} (1 - (1-x)^{m+1}) = -\ln(1-x),$$

we will write

$$g_m(x) = \frac{1}{m+1} (1 - (1-x)^{m+1}),$$

for all $x \in (0, 1)$ and for all m with $g_{-1}(x) = \lim_{m \to -1} g_m(x)$.

Substituting $F(x) = \int_0^x f(x) dx$ and $f(x)$ as given by (20.4), we obtain for $r = 1$,

$$f_{1,n,m,k}(x) = \frac{\gamma_1}{\sigma} e^{-\gamma_1 x/\sigma}, \quad x > 0, \ \sigma > 0. \qquad (20.6)$$

Thus,

$$X(1, n, m, k) \in E\left(\frac{\sigma}{\gamma_1}\right).$$

From (20.2), we obtain the moment generating function $M_{r,n,m,k}(t)$ of $X(r, n, m, k)$, when $X \in E(\sigma)$ as

$$M_{r,n,m,k}(t) = \int_0^\infty e^{tx} \frac{c_r}{(r-1)!} e^{-\gamma_r \sigma^{-1} x} \left\{ \frac{1}{m+1} \left[1 - e^{-(m-1)\gamma_r \sigma^{-1} x} \right] \right\}^{r-1} dx$$

$$= \frac{c_r}{(r-1)!} \int_0^\infty e^{-y(\gamma_r - t\sigma)} \left\{ \frac{1}{m+1} \left[1 - e^{-(m+1)y} \right] \right\}^{r-1} dy \qquad (20.7)$$

Using the following property (see Gradsheyn and Ryzhik (1965), p. 305)

$$\int_0^\infty e^{-ay}(1 - e^{-by})^{r-1} dy = \frac{1}{b} B\left(\frac{a}{b}, r\right) \quad \text{and} \quad \gamma_r + i(m+1) = \gamma_{r_i},$$

we obtain from (20.7)

$$M_{r,n,m,k}(t) = \frac{c_r}{(r-1)!} \frac{(r-1)!}{\prod\limits_{j=1}^{r} \gamma_j \left(1 - \frac{t\sigma}{\gamma_j}\right)} = \prod_{j=1}^{r} \left\{1 - \frac{t\sigma}{\gamma_j}\right\}^{-1}. \tag{20.8}$$

Thus,

$$X(r+1,n,m,k) \stackrel{d}{=} X(r,n,m,k) + \sigma \frac{W}{\gamma_{r+1}}. \tag{20.9}$$

It follows from (20.9) that

$$X(r+1,n,m,k) \stackrel{d}{=} \sigma \sum_{j=1}^{r+1} \frac{W_j}{\gamma_j}, \tag{20.10}$$

where $W_1, \ldots, W_{r+1}$ are i.i.d. with $W_i \in E(1)$.

If we take $k = 1$, $m = 0$, then we obtain from (20.10) the well-known result (see Ahsanullah and Nevzorov (2005, p. 51) for order statistics for $x \in E(\sigma)$:

$$X_{r+1,n} \stackrel{d}{=} \sigma \sum_{j=1}^{r+1} \frac{W_j}{n-j+1}. \tag{20.11}$$

From (20.9) it follows that $\gamma_{r+1}\{X(r+1,n,m,k) - X(r,n,m,k)\} \in E(\sigma)$. This property can also be obtained by considering the joint pdf of $X(r+1,n,m,k)$ and $X(r,n,m,k)$, using the transformations $U_1 = X(r,n,m,k)$ and $D_{r+1} = \gamma_{r+1}(X(r+1,n,m,k) - X(r,n,m,k))$ and integrating with respect to U_1.   □

Exercise 20.2. For the exponential distribution with $F(x) = 1 - e^{-x}$, $x \geqslant 0$, prove that $X(r,n,m,k) - X(r-1,n,m,k)$ and $X(r-1,n,m,k)$, $1 < r \leqslant n$, are independent.

Exercise 20.3. For the Pareto distribution with $F(x) = 1 - x^{-\delta}$, $x \geqslant 1$, $\delta > 0$, show that $E(X(r,n,m,k))^s = \gamma_1^* \cdots \gamma_r^*$, where $\gamma_i^* = k - \frac{s}{\delta} + (n-i)(m+1)$, $i = 1, \ldots, r$.

Example 20.3 (Lower Generalized Order Statistics (lgos)). Suppose $X^*(1,n,m,k)$, $X^*(2,n,m,k)$, ..., $X^*(n,n,m,k)$ are n lower generalized order statistics from an absolutely continuous cumulative distribution function (cdf) $F(x)$ with the corresponding probability density function (pdf) $f(x)$. Their joint pdf $f_{1,2,\ldots,n}^*(x_1, x_2, \ldots, x_n)$ is given by

$$f_{1,2,\ldots,n}^*(x_1, x_2, \ldots, x_n) = k \prod_{j=1}^{n-1} \gamma_j \prod_{i=1}^{n-1} (F(x_i))^m (F(x_n))^{k-1} f(x)$$

for $F^{-1}(1) \geqslant x_1 \geqslant x_2 \geqslant \cdots \geqslant F^{-1}(0)$, $m \geqslant -1$, where $\gamma_r = k + (n-r)(m+1)$, $r = 1, 2, \ldots, n-1$, $k \geqslant 1$, and n is a positive integer.

The marginal pdf of the r-th lower generalized order (lgos) statistics is

$$f^*_{r,n,m,k}(x) = \frac{c_r}{\Gamma(r)}(F(x))^{\gamma_r - 1}(g_m F(x))^{r-1} f(x), \tag{20.12}$$

where

$$c_r = \prod_{i=1}^{r} \gamma_i,$$

$$g_m(x) = \begin{cases} \dfrac{1}{m+1}\left(1 - x^{m+1}\right), & \text{for } m \neq -1, \\ -\ln x, & \text{for } m = -1, \end{cases}$$

Since $\lim_{m \to -1} g_m(x) = -\ln x$, we will take $g_m(x) = \frac{1}{m+1}(1 - x^{m+1})$ for all m with $g_{-1}(x) = -\ln x$.

For $m = 0$, $k = 1$, $X^*(r,n,m,k)$ reduces to the order statistics $X_{n-r+1,n}$ from the sample $X_1, X_2, \ldots, X_n$. If $m = -1$, then $X^*(r,n,m,k)$ reduces to the r-th lower k-record value.

Example 20.4. If $F(x)$ is absolutely continuous, then

$$\overline{F}^*_{r,n,m,k}(x) = 1 - F^*_{r,n,m,k}(x) = \begin{cases} I_{\alpha(x)}\left(r, \dfrac{\gamma_r}{m+1}\right), & \text{if } m > -1, \\ \Gamma_{\beta(x)}(r), & \text{if } m = -1, \end{cases}$$

where

$$\alpha(x) = 1 - (F(x))^{m+1}, \quad I_x(p,q) = \frac{1}{B(p,q)}\int_0^x u^{p-1}(1-u)^{q-1}du, \quad x \leqslant 1,$$

$$\beta(x) = -k\ln F(x), \quad \Gamma_x(r) = \frac{1}{\Gamma(r)}\int_0^x u^{r-1}e^{-u}du,$$

and

$$B(p,q) = \frac{\Gamma(p)\Gamma(q)}{\Gamma(p+q)}.$$

Proof. For $m > -1$ we get that

$$1 - F^*_{r,n,m,k}(x) = \frac{c_{r-1}}{\Gamma(r)}\int_x^\infty (F(u))^{\gamma_r - 1}(g_m(F(u)))^{r-1} f(u)du$$

$$= \frac{c_{r-1}}{\Gamma(r)}\int_x^\infty (F(u))^{\gamma_r - 1}\left[\frac{1 - (F(u))^{m+1}}{m+1}\right]^{r-1} f(u)du$$

$$= \frac{c_{r-1}}{\Gamma(r)}\frac{1}{(m+1)^r}\int_0^{1-(F(x))^{m+1}} t^{r-1}(1-t)^{(\gamma_r/(m+1))-1}dt$$

$$= I_{\alpha(x)}\left(r, \frac{\gamma_r}{m+1}\right).$$

For $m = -1$, $\gamma_j = k$, $j = 1, 2, \ldots, n$, and

$$1 - F^*_{r,n,m,k}(x) = \int_x^\infty \frac{k^r}{\Gamma(r)} (F(u))^{k-1} (-\ln F(u))^{r-1} f(u) du$$

$$= \int_0^{-k \ln F(x)} \frac{t^{r-1} e^{-t}}{\Gamma(r)} dt$$

$$= \Gamma_{\beta(x)}(r),$$

avec

$$\beta(x) = -k \ln F(x). \qquad \qquad \qquad \square$$

Example 20.5. Suppose that X is an absolutely continuous random variable with cdf $F(x)$ and pdf $f(x)$. It appears that the following relations are valid:

$$\gamma_{r+1} \left(F^*_{r+1,n,m,k}(x) - F^*_{r,n,m,k}(x) \right) = \frac{F(x)}{f(x)} f^*_{r+1,n,m,k}(x), \quad \text{for } m > -1,$$

and

$$k \left(F^*_{r+1,n,m,k}(x) - F^*_{r,n,m,k}(x) \right) = \frac{F(x)}{f(x)} f^*_{r+1,n,m,k}(x), \quad \text{for } m = -1,$$

Proof. For $m > -1$ the following equalities are valid:

$$F^*_{r+1,n,m,k}(x) - F^*_{r,n,m,k}(x) = I_{\alpha(x)} \left(r, \frac{\gamma_r}{m+1} \right) - I_{\alpha(x)} \left(r+1, \frac{\gamma_{r+1}}{m+1} \right)$$

$$= I_{\alpha(x)} \left(r, \frac{\gamma_r}{m+1} \right) - I_{\alpha(x)} \left(r+1, \frac{\gamma_r}{m+1} - 1 \right).$$

We know that

$$I_x(a,b) - I_x(a+1, b-1) = \frac{\Gamma(a+b)}{\Gamma(a+1)\Gamma(b)} x^a (1-x)^{b-1}.$$

Thus,

$$F^*_{r+1,n,m,k}(x) - F^*_{r,n,m,k}(x) = \frac{\Gamma\left(r + \frac{\gamma_r}{m+1}\right)}{\Gamma(r+1)\Gamma\left(\frac{\gamma_r}{m+1}\right)} \left(1 - (F(x))^{m+1}\right)^r (F(x)^{m+1})^{\frac{\gamma_r}{m+1} - 1}$$

$$= \frac{\gamma_1 \cdots \gamma_r}{\Gamma(r+1)} \left(\frac{1 - (F(x))^{m+1}}{m+1} \right)^r (F(x))^{\gamma_{r+1}}$$

$$= \frac{F(x)}{\gamma_{r+1} f(x)} f^*_{r+1,n,m,k}(x).$$

Hence,

$$\gamma_{r+1} \left(F^*_{r+1,n,m,k}(x) - F^*_{r,n,m,k}(x) \right) = f^*_{r+1,n,m,k}(x) \frac{F(x)}{f(x)}.$$

For $m = -1$ one gets that

$$F^*_{r+1,n,m,k}(x) - F^*_{r,n,m,k}(x) = \Gamma_{\beta(x)}(r) - \Gamma_{\beta(x)}(r+1),$$

where $\beta(x) = -k \ln F(x)$. Hence

$$F^*_{r+1,n,m,k}(x) - F^*_{r,n,m,k}(x) = (\beta(x))^r e^{-\beta(x)} \frac{1}{\Gamma(r+1)}$$

$$= \frac{(F(x))^k}{\Gamma(r+1)} (-k \ln F(x))^r.$$

Thus,

$$k\left[F^*_{r+1,n,m,k}(x) - F^*_{r,n,m,k}(x)\right] = \frac{F(x)}{f(x)} f^*_{r+1,n,m,k}(x). \qquad \square$$

Example 20.6 (lgos of power function distribution). Let X have a power function distribution with

$$F(x) = x^\delta, \quad 0 < x < 1, \quad \delta > 0.$$

Then

$$X^*(r+1,n,m,k) \stackrel{d}{=} X^*(r,n,m,k)W_{r+1},$$

where W_{r+1} is independent of $X^*(r,n,m,k)$ and the pdf of W_{r+1} is

$$f_{r+1}(w) = \delta\gamma_{r+1} w^{\delta\gamma_{r+1}-1}, \quad 0 < w < 1, \quad \delta > 0.$$

Proof. Let $Y = X^*(r,n,m,k)W_{r+1}$. Then for $m > -1$

$$F_Y(x) = P(Y \leqslant x) = P(X^*(r,n,m,k)W_{r+1} \leqslant x)$$

is given as

$$F_Y(x) = \int_0^x f^*_{r,n,m,k}(u)du + \int_x^1 \left(\frac{x}{u}\right)^{\delta\gamma_{r+1}} f^*_{r,n,m,k}(u)du.$$

Thus,

$$F_Y(x) = F^*_{r,n,m,k}(x) + \int_x^1 \left(\frac{x}{u}\right)^{\delta\gamma_{r+1}} f^*_{r,n,m,k}(u)du. \qquad (20.13)$$

Differentiating (20.13) with respect to x, we obtain

$$f_Y(x) = f^*_{r,n,m,k}(x) - f^*_{r,n,m,k}(x) + \int_x^1 \frac{\delta\gamma_{r+1}}{u^{\delta\gamma_{r+1}}} (x)^{\delta\gamma_{r+1}-1} f^*_{r,n,m,k}(u)du. \qquad (20.14)$$

We obtain on simplification that

$$\frac{f_Y(x)}{(x)^{\delta\gamma_{r+1}-1}} = \int_x^1 \frac{\delta\gamma_{r+1}}{u^{\delta\gamma_{r+1}}} f^*_{r,n,m,k}(u)du. \qquad (20.15)$$

Differentiating both sides of (20.15) with respect to x, we get the following equality:

$$\frac{f_Y'(x)}{(x)^{\delta\gamma_{r+1}-1}} - \frac{f_Y(x)}{(x)^{\delta\gamma_{r+1}}}(\delta\gamma_{r+1} - 1) = -\frac{\delta\gamma_{r+1}f_{r,n,m,k}^*(x)}{x^{\delta\gamma_{r+1}}}. \tag{20.16}$$

On simplification from (20.16), we obtain

$$f_Y'(x) - \frac{\delta\gamma_{r+1} - 1}{x}f_Y(x) = -\frac{\delta^2 c_r x^{\delta(\gamma_r - 1)}}{\Gamma(r)x}\left[\frac{1 - x^{\delta(m+1)}}{m+1}\right]^{r-1} x^{\delta-1}. \tag{20.17}$$

Multiplying both sides of (20.17) by $x^{-(\delta\gamma_{r+1}-1)}$, one gets that

$$\frac{d}{dx}\left(f_Y(x)x^{-(\delta\gamma_{r+1}-1)}\right) = -\frac{\delta^2 c_r x^{\delta(m+1)-1}}{\Gamma(r)}\left[\frac{1 - x^{\delta(m+1)}}{m+1}\right]^{r-1}.$$

Thus,

$$f_Y(x)x^{-(\delta\gamma_{r+1}-1)} = c - \int \frac{\delta^2 c_r x^{\delta(m+1)-1}}{\Gamma(r)}\left[\frac{1 - x^{\delta(m+1)}}{m+1}\right]^{r-1} dx$$

$$= c + \frac{\delta c_r}{\Gamma(r+1)}\left[\frac{1 - x^{\delta(m+1)}}{m+1}\right]^{r}, \tag{20.18}$$

where c is constant.

Hence

$$f_Y(x) = cx^{\delta\gamma_{r+1}-1} + \frac{\delta c_r x^{\delta\gamma_{r+1}-1}}{\Gamma(r+1)}\left[\frac{1 - x^{\delta(m+1)}}{m+1}\right]^{r}.$$

Since $f_Y(x)$ is a pdf with the conditions $F_Y(0) = 0$ and $F_Y(1) = 1$, we must have $c = 0$.
From the above equation, we obtain

$$f_Y(x) = \frac{\delta c_r x^{\delta\gamma_{r+1}-1}}{\Gamma(r+1)}\left[\frac{1 - x^{\delta(m+1)}}{m+1}\right]^{r}. \tag{20.19}$$

Thus, $Y = X^*(r+1, n, m, k)$.

The proof for the case $m = -1$ is similar. $\qquad\square$

Check your solutions

Exercise 20.1. Let $U = X(r-1, n, m, k)$ and $V = X(r, n, m, k) - X(r-1, n, m, k)$. Then the joint pdf, $f_{U,V}(u, v)$ of U and V is given as

$$f_{U,V}(u, v) = \frac{c_{r-1}}{\Gamma(r-1)}g_m^{r-2}(1 - e^{-u})e^{-\gamma_{r-1}u}e^{-\gamma_r v}, \quad 0 < u, \ v < \infty.$$

Thus, U and V are independent.

Exercise 20.2. Use the equality

$$F_{r,n,m,k}\left(r,\frac{\lambda_r}{m+1}\right) - F_{r,n,m,k}\left(r+1,\frac{\lambda_{r+1}}{m+1}\right) = \frac{1}{\lambda_{r+1}}\frac{\overline{F}(x)}{f(x)}f_{r+1,n,m,k}(x).$$

Then one gets for $m > -1$ that

$$F_{r,n,m,k}\left(r,\frac{\lambda_r}{m+1}\right) - F_{r,n,m,k}\left(r+1,\frac{\lambda_{r+1}}{m+1}\right) = I_{\alpha(x)}\left(r,\frac{\gamma_r}{m+1}\right) - I_{\alpha(x)}\left(r+1,\frac{\gamma_{r+1}}{m+1}\right).$$

Now, applying the relation

$$I_x(a,b) - I_x(a+1,b-1) = \frac{\Gamma(a+b)}{\Gamma(a+1)\Gamma(b)}x^a(1-x)^{b-1},$$

we obtain that

$$I_{\alpha(x)}\left(r,\frac{\gamma_r}{m+1}\right) - I_{\alpha(x)}\left(r+1,\frac{\gamma_r}{m+1}-1\right) = \frac{\Gamma\left(r+\frac{\gamma_r}{m+1}\right)}{\Gamma(r+1)\Gamma\left(\frac{\gamma_r}{m+1}\right)}(\alpha(x))^r(1-\alpha(x))^{\frac{\gamma_r}{m+1}-1}.$$

As it is known,

$$B\left(r,\frac{\gamma_r}{m+1}\right) = \frac{\Gamma(r)(m+1)^r}{c_r}.$$

Hence

$$\frac{\Gamma\left(r+\frac{\gamma_r}{m+1}\right)}{\Gamma(r+1)\Gamma\left(\frac{\gamma_r}{m+1}\right)}(\alpha(x))^r(1-\alpha(x))^{\frac{\gamma_r}{m+1}-1}$$

$$= \frac{c_r}{\Gamma(r+1)(m+1)^r}(1-(\overline{F}(x))^{m+1})^r(\overline{F}(x))^{m+1})^{\frac{\gamma_r}{m+1}-1}.$$

Thus,

$$F_{r,n,m,k}\left(r,\frac{\gamma_r}{m+1}\right) - F_{r,n,m,k}\left(r+1,\frac{\gamma_{r+1}}{m+1}\right) = \frac{1}{\gamma_{r+1}}\frac{\overline{F}(x)}{f(x)}f_{r+1,n,m,k}.$$

If $m = -1$ then

$$F_{r,n,m,k}\left(r,\frac{\gamma_r}{m+1}\right) - F_{r,n,m,k}\left(r+1,\frac{\gamma_{r+1}}{m+1}\right) = \Gamma_{\beta(x)}(r) - \Gamma_{\beta(x)}(r+1).$$

Using the relations

$$\Gamma_{\beta(x)}(r) - \Gamma_{\beta(x)}(r+1) = \frac{(\beta(x))^r e^{-\beta(x)}}{\Gamma(r+1)}$$

$$= \frac{(-k\ln\overline{F}(x))^r e^{k\ln\overline{F}(x)}}{\Gamma(r+1)}$$

$$= \frac{(-k\ln\overline{F}(x))^r(\overline{F}(x))^k}{\Gamma(r+1)},$$

we obtain that

$$F_{r,n,m,k}\left(r,\frac{\lambda_r}{m+1}\right) - F_{r,n,m,k}\left(r+1,\frac{\lambda_{r+1}}{m+1}\right) = \frac{1}{k}\frac{\overline{F}(x)}{f(x)}f_{r+1,n,m,k}(x).$$

Exercise 20.3 (hint). *Use the following equality:*

$$\int_1^\infty x^{-\delta(\gamma_r-1)}\left[\frac{1}{m+1}(1-x^{-\delta(m+1)})\right]^{r-1}\delta x^{-(\delta+1)}dx = \gamma_1\cdots\gamma_r,$$

where $\gamma_i = k+(n-i)(m+1)$, $i = 1,2,\ldots,r$.

Chapter 21

Compliments and Problems

In this chapter we will present some problems which will be helpful to study order statistics

21.1. Suppose we have observed the vector of order statistics $X_{r,n}, \ldots, X_{r,n}$, $1 \leqslant r \leqslant n$. *Show* that the BLUE of σ_1 and δ_2 are given by

$$\hat{\sigma}_1 = \frac{1}{2(s-r)} \left\{ (n-2r)X_{s,n} + (2s-n-1)X_{r,n} \right\}$$

$$\hat{\sigma}_2 = \frac{n+1}{s-r} \left(X_{s,n} - X_{r,n} \right)$$

21.2. Suppose $f_\sigma(x) = 1$, $\sigma \leqslant x \leqslant \sigma + 1$.

Show that BLUE of σ based on $X_{1,n}, \ldots, X_{n,n}$ is

$$\hat{\sigma} = \frac{X_{1,n} + X_{n,n} - 1}{2}.$$

21.3. Show that if for $1 < j \leqslant n$,

$$X_{j,n} - X_{j-1,n} \overset{d}{=} X_{1,n-j-1},$$

for two consecutive values of j, then

$$F(x) = 1 - e^{-x/\sigma}, \quad x \geqslant 0, \quad \sigma > 0,$$

$$F(n) = 1 - e^{-x/\sigma}, \quad x > u, \quad \sigma > u,$$

21.4. Let X be a discrete uniform $[1, n]$ random variable.

a. Determine μ_{12}

b. Determine the variances of $X_{1,3}$ and $X_{3,3}$.

21.5. Suppose that

$$E(X_{n,n} - X_{n-1,n}) = \frac{1}{n}, \quad n = 2, 3, \ldots$$

Determine the common distribution of X_i's.

M. Ahsanullah et al., *An Introduction to Order Statistics*,
Atlantis Studies in Probability and Statistics 3,
DOI: 10.2991/978-94-91216-83-1_21, © Atlantis Press 2013

21.6. Suppose x_1, x_2, and X_3 are *i.i.d.* positive and absolutely continuous random variables. Assume that

$$X_{2,3} - X_{1,3} \stackrel{d}{=} X_{1,2}$$
$$X_{3,3} - X_{1,3} = X_{2,2}.$$

Prove that X_i's are exponentially distributed.

21.7. Let X be a positive and bounded symmetric random variable having an absolutely continuous distribution function $F(x)$. Without any loss of generality we will assume that $\inf \{x \mid F(x) > 0\} = 0$ and $F(1) = 1$. Then the following two properties are equivalent:

a. X has the cdf $F(x) = x^{\alpha}$, $0 < x < 1$

b. X and $X_{1,n}/X_{2,n}$ are identically distributed for some $n \geqslant 2$ and F belongs to C_1. (F belongs to class C_1 if either $F(x_1 x_2) \geqslant F(x_1)F(x_2)$ or $F(x_1 x_2) \leqslant F(x_1)F(x_2)$). Ahsanullah (1989)

21.8. Let $U_1, U_2, \ldots, U_n$ be independent and identically distributed (iid) random variables (rv's) with a common distribution uniform on $[0, 1]$. Let $\{V_1, V_2, \ldots, V_n\}$ and $\{U_1, U_2, \ldots, U_n\}$ be iid random vectors. Show that

$$U_{k,n} \stackrel{d}{=} U_{k,m}V_{m+1,n}$$

Nevzorov (2001).

21.9. In recent years, some researchers have considered a generalized Pearson differential equation (GPE), given by

$$\frac{1}{f(x)}\frac{d}{dx}f(x) = \frac{\sum\limits_{j=0}^{m} a_j x^j}{\sum\limits_{j=0}^{n} b_j x^j},$$

where $m, n \in \mathbb{N} \setminus \{0\}$ and the coefficients a_j and b_j are real parameters. The system of continuous univariate *pdf*'s generated by GPE is called a generalized Pearson system, which includes a vast majority of continuous *pdf*'s, by proper choices of these parameters. Suppose that (x_r), $r = 1, 2, \ldots, n$ is a random sample of size n, that is, a sequence of n independent and identically distributed (*i.i.d.*) random variables (*rv's*) from a population with cumulative distribution function (cdf) $F_X(x)$ for a continuous random variable X. Suppose that $(X_{(r)})$, $r = 1, 2, \ldots, n$ be the corresponding order statistics. $F_{r,n}(x)$, $r = 1, 2, \ldots, n$ denotes the probability density function (pdf) of the r-th order statistic $X_{r,n}$, then it is defined as

$$f_{r,n}(x) = \frac{1}{B(r,n-r+1)}[F(x)]^{r-1}[1-F(x)]^{n-r}f(x),$$

where

$B(a,b) = \int_0^1 t^{a-1}(1-t)^{b-1}dt$ denotes the Beta function and

$I_p(a,b) = \frac{1}{B(a,b)}\int_0^p t^{a-1}(1-t)^{b-1}dt$ denotes the incomplete Beta function.

(a) Develop a generalized Pearson differential equation (GPE) satisfied by the probability density function (pdf) of the r-th order statistic $X_{(r,n)}$ if the underlying distributions are uniform, Pareto and exponential.

(b) Using the generalized Pearson differential equation developed in (a), determine the differential equations satisfied by the characteristic functions of the probability density function (pdf) of the r-th order statistic $X_{r,n}$ if the underlying distributions are uniform, Pareto and exponential. Hence, deduce the recurrence relations for moments.

(c) Using the generalized Pearson differential equation developed in (a), determine the differential equations satisfied by the cumulant generating function of the probability density function (pdf) of the r-th order statistic $X_{r,n}$ if the underlying distributions are uniform, Pareto and exponential. Hence, deduce the recurrence relations for cumulants

21.10. Suppose that x_r, $r = 1, 2, \ldots, n$ is a random sample of size n, that is, a sequence of n independent and identically distributed (*i.i.d.*) random variables (*rv's*) from a population with cumulative distribution function (cdf) $F_X(x)$ for a non-negative continuous random variable X. Suppose that $X_{(r)}$, $r = 1, 2, \ldots, n$ be the corresponding order statistics. Let $E(|X|) < \infty$. If $M_n = X_{(n)} = \text{Max}(x_r)$, $r = 1, 2, \ldots, n$, then show that

(i) $E(|X_{(r)}|) < \infty$.

(ii) $E(M_n) \leqslant E(M_{n-1}) + \int_0^\infty F^{n-1}(x)[1-F(x)]dx$, $n = 1, 2, 3, \ldots$.

(iii) Evaluate $E(M_n)$ if the underlying distributions are uniform, Pareto and exponential.

21.11. Suppose that $X_1, X_2, \ldots, x_n$ are n independent random sample from an absolutely continuous distribution with cdf $F(x)$ and pdf $f(x)$, Let $X_{1,n} \leqslant X_{2,n} \leqslant \cdots \leqslant X_{n,n}$ be the corresponding order statistics. If $M_n = X_{n,n}$ and if H_n denotes the Shannon entropy of M_n, then show that

$$H_n = 1 - \frac{1}{n} - \ln n - \int_{-\infty}^\infty f_{n,n}(x)\ln[f_{n,n}(x)]dx.$$

Hence show that

$$\frac{n-2}{2} \leqslant H_n + \Psi(n) + \int_{-\infty}^\infty f_{n,n}(x)\ln[f_{n,n}(x)]dx \leqslant \frac{2n-3}{2n},$$

where $\Psi(z)$ denotes the psi (digamma) function.

21.12.   Suppose that $X_1, X_2, \ldots, x_n$ are n independent random sample from an absolutely continuous distribution with cdf $F(x)$ and pdf $f(x)$, Let $X_{1,n} \leqslant X_{2,n} \leqslant \cdots \leqslant X_{n,n}$ be the corresponding order statistics. If $M_n = X_{n,n}$ and if H_n denotes the Shannon entropy of M_n, then show that

Let $S(x) = \sum_{r=1}^{n} f_{r,n}(x)$. If $f_X(x)$ denotes the pdf of the rv X, then express $S(x)$ as a linear convex combination of order statistics pdf's, that is, show that

$$S(x) = n f_X(x).$$

Bibliography

Abramowitz, M. and Stegan, I. (1972). *Handbook of Mathematical Functions.* Dover, New York, NY.

Aczel, J. (1966). *Chapters on Functional Equations and Their Applications.* Academic Press, New York, NY.

Adke, S. R. (1993). Records generated by Markov sequence. *Statistics and Probability Letters.* 18, 257–263.

Ahsanullah, M. (1975). A characterization of the exponential distribution. In G. P. Patil, S. Kotz and J. Ord eds. *Statistical Distributions in Scientific Work*, Vol. 3, 71–88. Dordecht-Holland , D. Reidel Publishing Company.

Ahsanullah, M. (1976). On a classification of the exponential distribution by order statistics, *J. Appl. Prob.* , 13, 818–822.

Ahsanullah, M. (1977). A Characteristic Property of the Exponential Distribution. *Ann. of Statist.* 5, 580–582.

Ahsanullah, M. (1978a). A characterization of the exponential distribution by spacings. *J. Appl. Prob.* , 15, 650–653.

Ahsanullah, M. (1978b). On characterizations of exponential distribution by spacings. *Ann. Inst. Stat. Math.* , 30, A. 429–433.

Ahsanullah, M. (1978c). Record Values and the Exponential Distribution. *Ann. Inst. Statist. Math.* 30, 429–433.

Ahsanullah, M. (1979). Characterization of the Exponential Distribution by Record Values. *Sankhya*, 41, B, 116–121.

Ahsanullah, M. (1980). Linear Prediction of Record Values for the Two Parameter Exponential Distribution. *Ann. Inst. Stat. Math.* , 32, A, 363–368.

Ahsanullah, M. (1981a). Record Values of the Exponentially Distributed Random Variables. *Statistiche Hefte*, 2, 121–127.

Ahsanullah, M. (1981b). On a Characterization of the Exponential Distribution by Weak Homoscedasticity of Record Values. *Biom. J.* , 23, 715–717.

Ahsanullah, M. (1981c). Record Values of the Exponentially Distributed Random Variables. *Statische Hefte*, 21, 121–127.

Ahsanullah, M. (1982). Characterizations of the Exponential Distribution by Some Properties of Record Values. *Statistiche Hefte*, 23, 326–332.

Ahsanullah, M. (1984). A characterization of the exponential distribution by higher order gap. *Metrika*, 31, 323–326.

Ahsanullah, M. (1986 a). Record Values from a Rectangular Distribution. *Pak. J. Stat.* , 2 (1), 1–6.

Ahsanullah, M. (1986 b). Estimation of the Parameters of a Rectangular Distribution by Record Values. *Comp. Stat. Quarterly*, 2, 119–125.

M. Ahsanullah et al., *An Introduction to Order Statistics*,
Atlantis Studies in Probability and Statistics 3,
DOI: 10.2991/978-94-91216-83-1, © Atlantis Press 2013

Ahsanullah, M. (1987a). Two Characterizations of the Exponential Distribution. *Comm. Statist. Theory Meth.* 16 (2), 375–381.

Ahsanullah, M. (1987b). Record Statistics and the Exponential Distribution. *Pak. J. Statist.* 3, A. 17–40.

Ahsanullah, M. and Holland, B. (1987). Distributional Properties of Record Values from the Geometric Distribution. *Statistica Neerlandica*, 41, 129–137.

Ahsanullah, M. (1988a). On a conjecture of Kakosian, Klebanov and Melamed. *Statische Hefte*, 29, 151–157.

Ahsanullah, M. (1988b). *Introduction to Record Statistics*, Ginn Press, Needham Heights, MA.

Ahsanullah, M. (1988c). Characteristic properties of order statistics based on a random sample size from an exponential distribution. *Statistica Neerlandicam;* 42, 193–197.

Ahsanullah, M. (1990). Estimation of the Parameters of the Gumbel Distribution Based on m Record Values. *Comp. Statist. Quarterly*, 3, 231–239.

Ahsanullah, M. (1991a). Some Characteristic Properties of the Record Values from the Exponential Distribution. *Sankhya* B, 53, 403–408.

Ahsanullah, M. (1991b). Record Values of the Lomax Distribution. *Statistica Neerlandica*, 45, 1, 21–29.

Ahsanullah, M. (1991c). On Record Values from the Generalized Pareto Distribution. *Pak. J. Stat.* , 7 (2) 129–136.

Ahsanullah, M. (1991d). Inference and Prediction Problems of the Gumbel Distributions based on Record Values. *Pak. J. Statist.* , 7#0 2, 53–62.

Ahsanullah, M. (1992a). Inference and Perdition Problems of the Generalized Pareto Distribution Based on Record Values. Non Parametric Statistics. Order Statistics and Non Parametrics. Theory and Applications, 49–57. Editors, P. K. Sen and I. A. Salama.

Ahsanullah, M. (1992b). Record Values of Independent and Identically Distributed Continuous Random Variables. *Pak. J. Statist.* , 8 (2), A, 9–34.

Ahsanullah, M (1994a). Records of the Generalized Extreme Value Distribution. *Pak. J Statist.* , 10 (1) A, 147–170.

Ahsanullah, M. (1994b). Record Values from Univariate Continuous Distributions. *Proceedings of the Extreme Value of the Extreme Value Theory and Applications*, 1–12.

Ahsanullah, M. (1994c). Records of Univariate Distributions. *Pak. J. Statist.* , 9 (3), 49–72.

Ahsanullah, M. (1994d). Some Inferences of Uniform Distribution based on Record Values. *Pak. J. Statist.* , 7, 27–33.

Ahsanullah, M. (1994e). Records Values, Random Record Models and Concomitants. *Journal of Statistical Research*, 28, 89–109.

Ahsanullah, M. (1995). Record Statistics. Nova Science Publishers Inc. New York, NY, USA.

Ahsanullah, M. (1996). Some Inferences of the Generalized Extreme Value Distribution Based on Record Values. *Journal of Mathematical Sciences*, 78 (1), 2–10.

Ahsanullah, M. (2000a). Concomitants of Record Values. *Pak. J. Statist.* , 16 (2), 207–215.

Ahsanullah, M. (2000b). Generalized order statistics from exponential distribution. *J. Statist. Plan. and Inf.* , 25, 85–91.

Ahsanullah, M. (2002). K-Record Values of the Type I Extreme Value Distribution. *Journal of Statistical Studies*, Special Volume, 283–292.

Ahsanullah, M. (2003a). Some characteristic properties of the generalized order statistics from exponential distribution. *J. Statist. Research*, 37 (2), 159–166.

Ahsanullah, M. (2003b). Some Inferences Based on K-Record Values of Uniform Distribution. *Stochastic Modelling and Applications*, 6 (1), 1–8.

Ahsanullah, M. (2004). *Record Values. Theory and Applications*, University Press of America, Lenham, Maryland. USA.

Ahsanullah, M. (2006). The generalized order statistics from exponential distribution. *Pak. J. Statist.*, 22 (2), 121–128.

Ahsanullah, M. , Berred, A. , and Nevzorov, V. B. (2011). On characterizations of the Exponential Distribution. *J. Appl. Statist. Sci.* , 19 (1), 37–42.

Ahsanullah, M. and Bhoj, D. (1996). Record Values of Extreme Value Distributions and a Test for Domain of Attraction of Type I Extreme Value Distribution (1996), *Sankhya*, 58, B, 151–158.

Ahsanullah, M. and Hamedani, G. G. (2010). Exponential distribution-Theory and Methods. Nova Science Publishers Inc. New York, NY. USA.

Ahsanullah, M. and Holland, B. (1984). Some Properties of the Distribution of Record Values from the Geometric Distribution. *Statische Hefte*, 25, 319–327.

Ahsanullah, M. and Holland, B. (1994). On the Use of Record Values to Estimate the Location and Scale Parameters of the Generalized Extreme Value Distribution. *Sankhya*, 56, A, 480–499.

Ahsanullah, M. and Kirmani, S. N. U. A. (1991). Characterizations of the Exponential Distribution by Lower Record Values. *Comm. Statist. Theory Meth.* , 20, (4), 1293–1299.

Ahsanullah, M. and Nevzorov, V. B. (2001). Ordered Random Variables. Nova Science Publishers Inc. , New York, NY, USA.

Ahsanullah, M. and Nevzorov, V. B. (2005). Order Statistics. Examples and Exercises, Nova Science Publishers,

Ahsanullah, M. and Nevzorov, V. B. (2011). Record Statistics. In *International Encyclopedia of Statistical Science*, 18, 1195–1202.

Ahsanullah, M. and Nevzorov, V. B. and Yanev, G. P. (2011). Characterizations of distributions via order statistics with random exponential shifts. *J. Appl. Statist. Sci.* , 18 (3), 297–317.

Ahsanullah, M. and Raqab, M. Z. (2000). Recurrence Relations for the Moment Generating Functions from Pareto and Gumbel Distribution. *Stochastic Modelling and Applications*, 2 (2), 35–48.

Ahsanullah, M. and Shakil, M. (2011). Record values of the ratio of two independent exponential random variables. *Journal of Statistical Theory and Applications*, 10 (3), 393–406.

Ahsanullah, M. and Wesolowski, J. (1998). Linearity of Best Non-adjacent Record Values, *Sankya*, B, 231–237.

Akhundov, I. , Balakrishnan, N. and Nevzorov, V. B. (2004). New characterizations by the properties by the parameters of midrange and related statistics. *Comm. in Statist. Theopry and Methods*, 33 (12), 3133–3143.

Alam, K. and Wallenius, K. T. (1979). Distribution of a sum of order statistics, *Scandinavian Journal of Statistics*, 6 (3), 123–126.

Akhundov, I. and Nevzorov, V. B. (2008). Characterizations of distributions via bivariate regression on differences of records. In *Records and Branching Processes*, Nova Science Publishers (eds. M. Ahsanullah and G. P. Yanev) 27–35.

Akhundov, I. and Nevzorov, V. B. (2009). On characterizations of t_2-distribution by regressional properties of maximums. *Metron International Journal of Statistics*, LXVII, 1, 51–56.

Ali, M. M. and Woo, J. (2010). Estimation of tail probability and reliability in exponentiated Pareto case. *Pak. J. Statist.* , 26 (1), 39–47.

Aliev, F. A. (1998). Characterizations of Discrete Distributions through Weak Records. *J. Appl. Satist. Sci.* , 8 (1), 13–16.

Aliev, F. A. (1999). New Characterizations of Discrete Distributions through Weak Records. *Theory Probab. Appl.* , 44 (4), 756–761.

Aliev, F. A. and Ahsanullah, M. (2002). On Characterizations of Discrete Distributions through Regressions of Record Values. *Pak. J. Statist.* , 18 (3), 415–421.

Aliev, F. A. (2003). A Comment on Unimodality of the distribution of Record Statistics. *Statistics & Probability Letters*, 64, 39–40.

Alpuim, T. M. (1985). Record Values in Populations with Increasing or Random Dimensions. *Metron*, 43 (3-4), 145–155.

AlZaid, A. A. and Ahsanullah, M. (2003). A Characterization of the Gumbell Distribution Based on Record Values. *Comm. Stat. Theory and Methods*, 32 (1), 2101–2108.

Arnold, B. C. (1983). Pareto Distributions. International Co-operative Publishing House. Fairland, Maryland.

Arnold, B. C. , Balakrishnan, N. , and Nagaraja, H. N. (1998). Records. John Wiley & Sons Inc. New York. NY, USA.

Azlarov, T. A. and Volodin, N. A. (1986). *Characterization Problems Associated with the Exponential Distribution*, Springer-Verlag, New York.

Bairamov, I. G. (1997). Some Distribution Free Properties of Statistics Based on Record Values and Characterizations of the Distributions through a Record. *J. Appl. Statist. Sci.* , 5 (1), 17–25.

Bairamov, I. G. and Ahsanullah, M. (2000). Distributional Relations between Order Statistics and the Sample itself and Characterizations of Exponential Distribution. *J. Appl. Stat. Sci.* , 10 (1), 1–16.

Bairamov, I. G. and Kotz, S. (2001). On Distributions of Exceedances Associated with Order Statistics and Record Values for Arbitrary Distributions. *Statistical Papers*, 42, 171–185.

Balakrishnan, N. , Ahsanullah, M. and Chan, P. S. (1992). Relations for Single and Product Moments of Record Values from Gumbel Distribution. *Stat. and Prob. Letters*, 15, 223–227.

Balakrishnan, N. , and Ahsanullah, M. and Chan, P. S. (1993) Recurrence Relations for Moments of Record Values from Generalized extreme value distribution. *Comm. Statist. Theory and Methods.* , 22 (5), 1471–1482.

Balakrishnan, N. , and Ahsanullah, M. (1993). Relations for Single and Product Moments of Record Values from Lomax Distribution. *Sankhya B*, 56 (2), 140–146.

Balakrishnan, N. , and Ahsanullah, M. (1994). Recurrence Relations for Single and Product Moments of Record Values from Generalized Pareto Distribution. *Commun. Statist. Theor. Meth.* , 23 (10), 2841–2852.

Balakrishnan, N. , and Ahsanullah, M. (1995). Relations for Single and Product Moments of Record Values from Exponential Distribution. *J. Apl. Statist. Sci.* , 2 (1), 73–88.

Balakrishnan, N. , Ahsanullah, M. and Chan, P. S. (1995). On the Logistic Record Values and Associated Inference. *J. Appl. Statist. Sci.* , 2, 233-248.

Balakrishnan, N. , and Balasubramanian, K. (1995). A Characterization of Geometric Distribution Based on Record Values. *J. Appl. Statist. Sci.* , 2, 277-282.

Balakrishnan, N. , and Nevzorov, V. B. (1998). Record of Records. In *Handbook of Statistics*, 16. (eds. N. Balakrishnan, C. R. Rao). Amsterdam, North Holland, 515–570.

Balakrishnan, N. , and Nevzorov,V. B. (2006). Record Values and record statistics. *Encyclopedia of Statistical Science.* The second edition. Wiley Interscience. 10, 6995–7006.

Ballerini, R. and Resnick, S. I. (1985). Records from Improving Populations. *J. Appl. Prob.* , 22, 487–502.

Ballerini, R. and Resnick, S. I. (1987). Records in the Presence of a Linear Trend. *Adv. in Appl. Prob.* , 19, 801–828.

Barton, D. E. and Mallows, C. L. (1965). Some Aspects of the Random Sequence. *Ann. Math. Statist,* , 36, 236–260.

Basak, P. (1996). Lower Record Values and Characterization of Exponential Distribution. *Calcutta Statistical Association Bulletin*, 46, 1–7.

Berred,A. and Nevzorov,V. B. (2009). Samples without replacement. : one property of distribution of range. *Notes of Sci. Semin, POMI*, 368, 53–58.

Blaquez, F. L. , Rebollo, J. L. (1997). A Characterization of Distributions Based on Linear Regression of Order Statistics and Record Values. *Sankhya*, 59, A, 311–323.

Blom, G. Thorburn, D. and Vessey, T. (1990). The Distribution of the Record Position and its Applications. *Amer. Statistician*, 44, 152–153.

Bruss, F. T. (1988). Invariant Record Processes and Applications to Best Choice Modeling. *Stoch. Proc. Appl.* , 17, 331–338.

Bunge, J. A. and Nagaraja, H. N. (1992). Exact Distribution Theory for Some Point Process Record Models. *Adv. Appl. Prob.* , 24, 20–44.

Chandler, K. N. (1952). The Distribution and Frequency of Record Values. *J. R. Statist. Soc.* B 14, 220–228.

Cinlar, E. (1975). *Introduction to Stochastic Processes.* Prentice-Hall, New Jersey, 1975.

Dallas, A. C. (1981). Record Values and the Exponential Distribution. *J. Appl. Prob.* , 18, 959–951.

Dallas,A. C. and Resnick, S. I. (1981). A Characterization Using Conditional Variance. *Metrika*, 28, 151–153.

David, H. A. (1981). Order Statistics. Second Edition. John Wiley and Sons, Inc. , New York, NY.

DeHaan, L. and Resnick, S. I. (1973). Almost Sure Limit Points of Record Values. *J. Appl. Prob.* , 10, 528–542.

Deheuvels, P. (1983). The Complete Characterization of the Upper and Lower Class of Record and Inter - Record Times of i. i. d. Sequence. *Zeit. Wahrscheinlichkeits theorie Verw. Geb.* , 62, 1–6.

Deheuvels, P. (1984). The Characterization of Distributions by Order Statistics and Record Values - a Unified Approach. *J. Appl. Prob.* , 21, 326–334.

Deheuvels, P. and Nevzorov, V. B. (1994). Limit Laws for k-record Times. *J. Stat. Plan. and Inf.* , 38 (3), 279–308.

Deheuvels, P. and Nevzorov, V. B. (1999). Bootstrap for maxima and records. *Notes of Sci. Seminars of POMI*, 260, 119–129.

Deken, J. G. (1978). Record Values, Scheduled Maxima Sequences. *J. Appl. Prob.* 15, 491–496.

Dembinska, A. and Wesolowski, J. (2003). Constancy of Regression for size two Record Spacings. *Pak. J. Statist.* , 19 (1), 143–149.

Dembinska, A. and Wesolowski, J. (2000). Linearity of Prediction for Non-adjacent Record Values. *J. Statist. Plann. Infer.* , 90, 195–205.

Dunsmore, J. R. (1983). The Future Occurrence of Records. *Ann. Inst. Stat. Math.* , 35, 267–277.

Dwass, M. (1964). External Processes. *Ann. Math. Statist.* , 35, 1718–1725.

Embrechts, P. and Omey, E. (1983). On Subordinated Distributions and Random Record Processes. *Ma. P. Cam. Ph.* , 93, 339–353.

Feller, W. (1966). *An Introduction to Probability Theory and its Applications.* Vol. II, Wiley, New York.

Foster, F. G. and Teichroew, D. (1955). A Sampling Experiment on the Powers of the Record Tests for a Trend in a Time Series. *J. R. Statist. Soc. Ser.* B, 17, 115–121.

Franco, M. and Ruiz, J. M. (2001). On Characterizations of Distributions by Expected Values of Order Statistics and Record Values with Gap. *Metrika*, 45, 107–119.

Franco, M. and Ruiz, J. M. (2001). Characterization of Discrete distributions based on conditional expectations of Record Values. *Statistical Papers*, 42, 101–110.

Fréchet, M. (1927). Sur La loi probabilité de l'écart Maximum. *Ann. de la Soc. Polonaise de Math.* , 6, 93–116.

Freudenberg, W. and Szynal, D. (1976). Limit Laws for a Random Number of Record Values. Bull. Acad. Polon. Sci. Ser. Math. Astr. Phys. 24, 195–199.

Galambos, J. (1987). *The Asymptotic Theory of Extreme Order Statistics.* Robert E. Krieger Publishing Co. Malabar, Florida.

Galambos, J. and Kotz, S. (1978). *Characterizations of Probability Distributions.* Lecture Notes in Mathematics. No. 675, Springer Verlag, New York.

Galambos, J. and Seneta, E. (1975). Record Times. *Proc. Amer. Math. Soc.* , 50, 383–387.

Galambos, J. and Simonelli, I, (1996). *Bonferroni-type Inequalities with Applications.* Springer, New York, NY, USA.

Gajek, L. and Gather, U. (1989). *Characterizations of the Exponential Distribution by Failure Rate and Moment Properties of Order Statistics*. Lecture Notes in Statistics, 51. Extreme Value Theory, Proceedings, 114–124. J. Husler, R. D. Reiss (Eds.), Springer-Verlag, Berlin, Germany.

Gaver, D. P. (1976). Random Record Models. *J. Appl. Prob.* , 13, 538–547.

Gaver, D. P. and Jacobs, P. A. (1978). Non Homogeneously Paced Random Records and Associated Extremal Processes. *J. Appl. Prob.* , 15, 543–551.

Garg, M. (2009). On generalized order statistics from Kumaraswamy distribution, *Tamsui Oxford Journal of Mathematical Sciences*, 25 (2), 153–166.

Glick N. (1978). Breaking Records and Breaking Boards. *Amer. Math. Monthly*, 85 (1), 2–26.

Gnedenko, B. (1943). Sur la Distribution Limite du Terme Maximum d'une Serie Aletoise. *Ann. Math.* , 44, 423–453.

Goldburger, A. S. (1962). Best Linear Unbiased Predictors in the Generalized Linear Regression Model. *J. Amer. Statist. Assoc.* , 57, 369–375.

Goldie, C. M. (1989). Records, Permutations and Greatest Convex Minorants. *Math Proc. Camb, Phil. Soc.* , 106, 169–177.

Goldie, C. M and Resnick, S. I. (1989). Records in Partially Ordered Set. *Ann. Prob.* , 17, 675–689.

Goldie, C. M. and Resnick, S. I. (1995). Many Multivariate Records. *Stoch. Proc. And Their Appl.* , 59, 185–216.

Goldie, C. M. and Rogers, L. C. G. (1984). The *k*-Record Processes are i. i. d. *Zeit. Wahr. Verw. Geb.* , 67, 197–211.

Gradshteyn, I. S. and Ryzhik, I. M. (1980). *Tables of Integrals, Series, and Products*, Corrected and Enlarged Edition. Academic Press, Inc.

Grosswald, E. and Kotz, S. (1981). An Integrated Lack of Memory Property of the Exponential Distribution. *Ann. Inst. Statist. Math.* , 33, A, 205–214.

Grudzien, Z. and Szynal, D. (1985). On the Expected Values of kth Record Values and Characterizations of Distributions. Probability and Statistical Decision Theory, Vol. A. (F. Konecny, J. Mogyorodi and W. Wertz, Eds.) Dordrecht - Reidel, 1195–214.

Gulati, S. and Padgett, W. J. , (1994a). Nonparametric Quantitle Estimation from Record Breaking Data. *Aust. J. of Stat.* , 36, 211–223.

Gulati, S. and Padgett, W. J. , (1994b). Smooth Nonparametric Estimation of Distribution and Density Function from Record Breaking Data. *Comm. In Stat. –Theory and Methods.* , 23, 1259–1274.

Gulati, S. and Padgett, W. J. , (1994c). Smooth nonparametric Estimation of the Hazard and Hazard Rate Functions from Record Breaking Data. *J. Stat. Plan. and Inf.* , 42, 331–341.

Gumbel, E. J. (1958). *Statistics of Extremes.* Columbia Univ. Press, New York.

Gumbel, E. J. (1960). Bivariate Exponential Distributions, *JASA*, 55, 698–707.

Gupta, R. C. (1984). Relationships between Order Statistic and Record Values and Some Characterization Results. *J. Appl. Prob.* 21, 425–430.

Gupta, R. C. and Kirmani, S. N. U. A. (1988). Closures and Monotonicity Properties of Non-Homogenious Poisson Processes and Record Values. *Probability in Engineering and Informational Sciences*, 2, 475–484.

Gupta, S. S. , Pillai, K. C. S. , and Steck, G. P. (1964). The distribution of linear functions of ordered correlated normal random variables with emphasis on range, *Biometrika*, 51, 143–151.

Gut, A. (1990 a). Converge rates for Record Times and the Associated Covering Processes. *Stoch. Processes Appl.* , 36, 135–152.

Gut, A. (1990 b). Limit Theorems for Record Times. Probability Theory and Mathematical Statistics. V. 1. (Proceedings of the 5th Vilanius Conference on Probability Theory and Mathematical Statistics) VSP/Mokslas, 490–503.

Guthrie, G. L. and Holmes, P. T. (1975). On Record and Inter-Record Times of a Sequence of Random Variables Defined on a Markov Chain. *Adv. Appl. Prob.* , 7, 195–214.

Haghighi-Tabab, D. and Wright, E. (1973). On the Distribution of Records in a Finite Sequence of Observations, with an Application to a Road Traffic Problem. *J. Appl. Prob.* , 10, 56–571.

Haiman, G. (1987). Almost Sure Asymptotic Behavior of the Record and Record Time Sequences of a Stationary Time Series. New Perspective in Theoretical and Applied Statistics, M. L. Puri, J. P. Vilaplana and W. Wertz, Eds.) John Wiley, New York, 459–466.

Haiman, G. Mayeur, N. , and Nevzorov, V. B. and Puri, M. L. (1998). Records and 2-Block Records of 1- dependent Stationary Sequences under Local Dependence. *Ann. Inst. Henri Poincaré*, 34, 481–503.

Haiman, G. and Nevzorov, V. B. (1995). Stochastic Ordering of the Number of Records. Stochastic Theory and Applications. Papers in Honor of H. A. David (H. N. Nagaraja, P. K. Sen and D. F. Morison, eds.) Springer Verlag, Berlin, 105–116.

Hall, P. (1975). Representations and Limit Theorems for Extreme Value Distributions. *J. Appl. Prob.* , 15, 639–644.

Hill, B. M. (1975). A Simple General Approach to Inference about the Tail of a Distribution. *Ann. statist.* , 3, 1163–1174.

Holmes, P. T. and Strawderman, W. (1969). A Note on the Waiting Times between Record Observations. *J. Appl. Prob.* , 6, 711–714.

Horwitz, J. (1980). Extreme Values from a Non Stationary Stochastic Process: an Application to Air Quality Analysis (with discussion). *Technometrics*, 22, 469–482.

Huang, Wen-Jang and Li, Shun-Hwa (1993). Characterization Results Based on Record Values. *Statistica Sinica*, 3, 583–599.

Jenkinson, A. F. (1955). The Frequency Distribution of the Annual Maximum (or mimimum) values of Meteorological Elements. *Quart. J. Meter. Soc.* , 87, 158–171.

Johnson, N. L. and Kotz, S. (1977). *Distributions in statistics: Continuous Multivariate Distributions*. Wiley, New York.

Kakosyan, A. V. , Klebanov, L. B. and Melamed, J. A. (1984). *Characterization of Distribution by the Method of Intensively Monotone Operators*. Lecture Notes Math. 1088, Springer Verlag, New York, N. Y.

Karlin, S. (1966). *A First Course in Stochastic Processes*. Academic Press, New York.

Kirmani, S. N. U. A. and Beg, M. I. (1983). On Characterization of Distributions by Expected Records. *Sankhya*, 48 A, 463–465.

Klebanov, L. B. and Melamed, J. A. (1983). A Method Associated with Characterizations of the exponential distribution. *Ann. Inst. Statist. Math.* , 35, A, 105–114.

Korwar, R. M. (1984). On Characterizing Distributions for which the Second Record has a Linear Regression on the First. *Sankhya* B 46, 108–109.

Korwar, R. M. (1990). Some Partial Ordering Results on Record Values. *Commun. Statist. -Thero. Meth.* , 19 (1), 299–306.

Lau, Ka-sing and Rao, C. R. (1982). Integrated Cauchy Functional Equation and Characterization of the Exponential. *Sankhya* a, 44, 72–90.

Leadbetter, M. R. , Lindgreen, G. and Rootzen, H. (1983). Extremes and Related Properties of Random Sequences and Series, Springer- Verlag, New York, N. Y.

Lin, G. D. (1987). On Characterizations of Distributions via Moments of Record Values. *Probab. Th. Rel. Fields*, 74, 479–483.

Lloyd, E. H. (1952). Least Squares Estimation of Location and Scale Parameters Using Order Statistics. *Biometrika*, 39, 88–95.

Malik, H. and Trudel, R. (1976). Distributions of product and quotient of order statistics, *University of Guelph Statistical Series*, 1975. 30, 1–25.

Malov, S. V. (1997). Sequential τ-ranks. *J. Appl. Statist. Sci.* , 5, 211–224.

Mann, N. R. (1969). Optimum Estimators for Linear Functions of Location and Scale Parameters. *Ann. Math. Statist.* , 40, 2149–2155.

Marsglia, G. and Tubilla, A. (1975). A Note on the Lack of Memory Pproperty of the Exponential Distribution. *Ann. Prob.* , 3, 352–354.

Mellon, B. (1988). *The Olympic Record Book.* Garland Publishing, Inc. New York, N. Y.

Mohan, N. R. and Nayak, S. S. (1982). A Characterization Based on the Equidistribution of the First Two Spacings of Record Values. *Z. Wahr. Verw. Geb.* 60, 219–221.

Morrison, M. and Tobias, F. (1965). Some statistical characteristics of a peak to average ratio. Technometrics, 7, 379–385.

Nadarajah, S. (2010). Distribution properties and estimation of the ratio of independent Weibull random variables. *AStA Advances in Statistical Analysis*, 94, 231–246.

Nadarajah, S. and Dey, D. K. (2006). On the product and ratio of t random variables. *Applied Mathematics Letters*, 19, 45–55.

Nadarajah, S. and Kotz, S. (2007). On the product and ratio of t and Bessel random variables. *Bulletin of the Institute of Mathematics*, 2 (1), 55–66.

Nagaraja, H. N. (1977). On a Characterization Based on Record Values. *Austral. J. Statist.* 19, 70–73.

Nagaraja, H. N. (1978). On the Expected Values of Record Values. *Austral. J. Statist.* , 20, 176–182.

Nagaraja, H. N. (1988) Record Values and Related Statistics – a Review, *Commun. Statist. Theory Meth.* , 17, 2223–2238.

Nagaraja, H. N. and Nevzorov, V. B. (1977). On characterizations Based on Record Values and Order Statistics. *J. Stat. Plan. and Inf.* , 61, 271–284.

Nagaraja, H. N. and Nevzorov, V. B. (1996). Correlations between Functions of the Records can be negative. *Stat. & Proab. Letters*, 29, 95–100.

Nagaraja, H. N. Sen, P. and Srivastava, R. C. (1989). Some Characterizations of Geometric Tail Distributions Based on Record Values. *Statistical Papers* 30, 147–155.

Nayak, S. S. (1981). Characterizations Based on Record Values. *J. Indian Statist. Assn.* , 19, 123–127.

Nayak, S. S. (1985). Record Values for and Partial Maxima of a Dependent Sequence. *J. Indian Statist. Assn.* , 23, 109–125.

Neuts, M. F. (1967). Waiting Times Between Record Observations. *J. Appl. Prob.* , 4, 206–208.

Nevzorova, L. N and Nevzorov,V. B. (1999). Ordered Random Variables. *Acta Applicandoe Mathematicae*, 58, 217–2229.

Nevzorova, L. N. , Nevzorov, V. B. and Balakrishnan, N. , (1997). Characterizations of distributions by extreme and records in Archimedian copula processes. In *Advances in the Theory and Pratice of Statistics - A volume in Honor of Samuel Kotz* (Eds. N. L. Johnson and N. Balakrishnan), pp. 469–478. John Wiley and Sons, New York, USA.

Nevzorov, V. B. (1988). Records. *Theo. Prob. Apl.* , 32, 201–228.

Nevzorov, V. B. (1992a). A Characterization of Exponential Distributions by Correlation between Records. *Mathematical Methods of Statistics*, 1, 49–54.

Nevzorov, V. B (1992b). A Characterization of Exponential Distributions by Correlation between the Records. *Mathematical Methods of Statistics*, 1, 49–54.

Nevzorov, V. B. (1995). Asymptotic distributions of Records in Non–stationary Schemes. *J. Stat. Plan. and Inf.* , 45, 261–273.

Nevzorov, V. B. (2001). *Records: Mathematical Theory. Translation of Mathematical Monographs*, Volume 194. American Mathematical Society. Providence, RI, USA

Nevzorov, V. B. (2004). Record models for sport data. In *Longgevity, ageing and degradation models.* (Eds. V. Antonov, C. Huber and M. Nikulin). St. Petersburg, 1, 198–200.

Nevzorov, V. B. (2007). Extremes and Student's t-distribution. *Probability and Statistics.* 12 (*Notes of Sci. Semin. POMI.*) 351, 232–237 (in Russian).

Nevzorov, V. B. and Ahsanullah, M. (2001a). Some Distributions of Induced Records. *Biometrical Journal*, 42 (8), 1069–1081.

Nevzorov, V. B. and Ahsanullah, M. (2001b). Extremes and Records for Concomitants of Order Statistics and Record Values. *J. Appli. Statist. Sci.* , 10 (3), 181–190.

Nevzorov, V. B. and Sagateljan,V. K. (2008). Characterizations of distributions by equalities of order statistics. In *Recent Developments in Order Random Varaables. Nova Science Publishers*, (eds. M. Ahsanullah and M. Z. Raqab). 67–76.

Nevzorov, V. B. and Zhukova,E. (1996). Wiener Process and Order Statistics. *J. Appl. Statist. Sci.* , 3 (4), 317–323.

Oncei, S. Y. , Ahsanullah, M,. Gebizlioglu and Aliev, F. A. (2001). Characterization of Geometric Distribution through Normal and Weak Record Values. *Stochastc Modelling and Applications*, 4 (1), 53–58.

Pawlas, P. and Szynal, D. (1999). Recurrence Relations for Single and Product Moments of K-th Record Values From Pareto, Generalized Pareo and Burr Distributions. *Commun. Statist. - Theory and Methods*, 28 (7), 1699–1701.

Pfeifer, D. (1981). Asymptotic Expansions for the Mean and Variance of D Logarithmic Inter-Record Times. *Meth. Operat. Res.* , 39, 113–121.

Pfeifer, D. (1982). Characterizations of Exponential Distributions by Independent Non Stationary Increments. *J. Appl. Prob.* , 19, 127–135.

Pfeifer, D. (1988). Limit Laws for Inter-Record Times for Non Homogeneous Markov Chains. *J. Organizational Behav. Statist.* , 1, 69–74.

Pickands III, J. (1971). The Two Dimensional Poisson Process and Extremal Processes. *J. Appl. Prob.* , 8, 745–756.

Rahimov, I. (1995). Record Values of a Family of Branching Processes. *IMA Volumes in Mathematics and its Application.* 84, 285–295.

Rahimov, I. and Ahsanullah, (2001). M. Records Generated by the total Progeny of Branching Processes. *Far East J. Theo. Stat*, 5 (10), 81–84.

Rahimov, I. and Ahsanullah, (2003). Records Related to Sequence of Branching Stochastic Process. *Pak. J. Statist.* 19, 73–98.

Ramachandran, B. and Lau, R. S. (1991). Functional Equations in Probability Theory. Academic Press, Boston, MA, USA.

Rao, C. R. (1983). An Extension of Denny's Theorem and its Application to Characterizations of Probability Distributions. A Festschrift for Erich L. Lehman. Ed. Bickel *et al.* Wordsworth International Group, Belmont, CA.

Raqab, M. Z. (1997). Bounds Based on Greatest Convex Minorants for Moments of Record Values. *Statistics and Probability Letters*, 36, 35–41.

Raqab, M. Z. and Amin, W. A. (1997). A note on reliability properties of k–record statistics. *Metrika*, 46, 245–251.

Raqab, M. Z. (1998). Order statistics from the Burr type X model. *Computers Mathematics and Applications*, 36, 111–120.

Raqab, M. Z. and Ahsanullah, M. , (2000). Relations for marginal and joint moment generating functions of record values from power function distribution. *J. Appl. Statist. Sci.* , 10 (1), 27–36.

Raqab, M. Z. (2002). Characterizations of Distributions Based on Conditional Expectations of Record Values. *Statistics and Decisions.* 20, 309–319.

Raqab, M. Z. and Ahsanullah, M. (2003). On Moment Generating Functions of Records from Extreme Value Distributions. *Pak. J. Statit.* 19 (1), 1–13.

Reiss, R. D. (1989). Approximate Distributions of Order Statistics. Springer-Verlag, New York, N. Y.

Renyi, A. (1962). Théorie des Éléments Saillants d'une Suite d'observations Collq. *Combinatorial Meth. Prob. Theory*, Aarhus University 104–115.

Renyi, A. (1973). Record Values and Maxima. Ann. Probab. 1, 650–662.

Resnick, S. (1987). Extreme Values, Regular Variation and Point Processes. Springer Verlag, New York.

Resnick, S. (1973). Record Values and Maxima. *Ann. Probab.* 1, 650–662.

Roberts, E. M. (1979). Revew of Statistics of Extreme Values with Application to Air Quality Data. Part II: application. *J. Air Polution Control Assoc.* 29, 733–740.

Salamingo, F. J. and Whitaker, L. D. (1986). On Estimating Population Characteristics from Record Breaking Observations. I. Parametric Results. *Naval Res. Log. Quart.* 25, 531–543.

Shah, B. K. (1970). Note on Moments of a Logistic Order Statistics, *Ann. Math. Statist.* 41, 2151–2152.

Shakil, M. and Ahsanullah, M. (2011). Record values of the ratio of Rayleigh random variables, *Pakistan Journal of Statistics*, 27 (3), 307–325.

Shakil, M. and Ahsanullah, M. (2012). Review on Order Statistics and Record Values from F^α Distributions, *Pakistan Journal of Statistics and Operation Research*, Vol. VIII, No. 1, pp. 101–120.

Shakil, M. , Kibria, B. M. G. , and Chang, K. -C. (2008). Distributions of the product and ratio of Maxwell and Rayleigh random variables. *Statistical Papers*, 49, 729–747.

Sen, P. K. On Moments of the Sample Quantiles. *Calcutta Statist. Assoc. Bull.* 9, 1–19.

Sen, P. K. (1973). Record Values and Maxima. Ann. *Probab.* 1, 650–662.

Sen, P. K. (1970). A Note on Order Statistics for Heterogeneous Distributions. *Ann. Math. Statist.* , 41, 2137–2139.

Shannon, C. E. (1948). A Mathematical Theory of Communication (concluded). *Bell. Syst. Tech. J.* , 27, 629–631.

Shawky, A. I. , and Abu-Zinadah, H. H. (2008). Characterizations of the exponentiated Pareto distribution based on record values. *Applied Mathematical Sciences*, 2 (26), 1283–1290.

Shawky, A. I. , and Abu-Zinadah, H. H. (2009). Exponentiated Pareto distribution: different method of estimations. *Int. J. Contemp. Math. Sci.* , 4 (14), 677–693.

Shawky, A. I. , and Bakoban, R. A. (2008). Characterizations from exponentiated gamma distribution based on record values, *JSTA*, 7 (3), 263–278.

Sethuramann, J. (1965). On a Characterization of the Three Limiting Types of Extremes. *Sankhya*, A 27, 357–364.

Shorrock, S. (1972). Record Values and Record Times. *J. Appl. Prob.* 9, 316–326.

Shorrock, S. (1973). Record Values and Maxima. *Ann. Probab.* 1, 650–662.

Shorrock, S. (1973) Record Values and Inter-Record Times. *J. Appl. Prob.* 10, 543–555.

Siddiqui, M. M. and Biondini, R. W. (1975). The Joint Distribution of Record Values and Inter-Record Times. *Ann. Prob.* 3, 1012–1013.

Singpurwala, N. D. (1972). Extreme Values from a Lognormal Law with Applications to Air Population Problems. *Technometrics* 14, 703–711.

Smith, R. L. (1986). Extreme Value Theory Based on the r Largest Annual Events. *J. Hydrology.* 86, 27–43.

Smith, R. L. and Weissman, I. (1987). Large Deviations of Tail Estimators based on the Pareto Approximations. *J. Appl. Prob.* 24, 619–630.

Springer, M. D. (1979). *The Algebra of Random Variables*, New York, John Wiley & Sons, Inc.

Srivastava, R. C. (1978). Two Characterizations of the Geometric Distribution by Record Values. *Sankhya* B, 40, 276–278.

Stam, A. J. (1985). Independent Poisson Processes by Record Values and Inter-Record Times. *Stoch. Proc. and Their Appl.* 19, 315–325.

Stepanov, A. V. (1990). Characterizations of a Geometric Class of Distributions. *Theory of Probability and Mathematical Statistics*, 41, 133–136 (English Translation)

Stepanov, A. V. (1992). Limit Theorems for Weak Records. *Theory of Probab. and Appl.* , 37, 579–574 (English Translation).

Stepanov, A. V. (1994). A Characterization Theorem for Weak Records. *Theory Prob. Appl.* , 38, 762–764 (English Translation).

Stepanov, A. V. (2001). Records when the Last Point of Increase is an Atom. J. *Appl. Statist. Sci.* 19, 2, 161–167.

Stepanov, A. V. (2003). Conditional Moments of Record Times. *Statistical Papers*, 44, 131–140.

Stigler, S. M. (1969). Linear functions of order statistics, *The Annals of Mathematical Statistics*, 40 (3), 770–788.

Stigler, S. M. (1974). Linear functions of order statistics with smooth weight functions, *Annals of Statistics*, 2, 676–693.

Strawderman, W. and Holmes, P. T. (1970). On the Law of Iterated Logarithm for Inter Record Times. *J. Appl. Prob.* , 7, 432–439.

Subrahmanian, K. (1970). On some applications of Mellin transforms to statistics: Dependent random variables, *SIAM J. Appl. Math.* , 19 (4), 658–662.

Tallie, C. (1981). A Note on Srivastava's Characterization of the Exponential Distribution Based on Record Values. *Statistical Distributions in Scientific Work.* Ed. C. Tallie, G. P. Patil and B. Boldessari, Reidel, Dordrecht, 4, 417–418.

Tata, M. N. (1969). On Outstanding Values in a Sequence of Random Variables. *Zeit. Wehrschein-lichkeitsth.* , 12, 9–20.

Teugels, J. L. (1984). On Successive Record Values in a Sequence of Independent and Identically Distributed Random Variables. *Statistical Extremes and Applications.* (Tiago de Oliveira, Ed.). Dordrecht-Reidel, 639–650.

Tiago de Oliveira, J. (1958). Extremal Distributions. *Rev. Fac. Cienc. Univ. Lisboa A*, 7, 215–227.

Tippett, L. H. C. (1925). On the Extreme Individuals and the Range of Samples Taken from a Normal Population. *Biometrika*, 17, 364–387.

Van der Vaart, H. R. (1961). Some extensions of the idea of bias. *Ann. Math. Statist.* , 32, 436–447.

Vervaat, W. (1973). Limit Theorems for Records from Discrete Distributions. *Stoch. Proc. and Their Appl.* , 1, 317–334.

Weissman, I. (1978). Estimations of Parameters and Large Quantiles Based on the k Largest Observations. *J. Amer. Stat. Assn.* , 73, 812–815.

Wesolowski, J. , and Ahsanullah, M. (2000). Linearity of Regression for Non-adjacent weak Records. *Statistica sinica*, 11, 30–52.

Westcott, M. (1977). The Random Record Model. *Proc. Roy. Soc. Lon.* , A. 356, 529–547.

Westcott, M. (1979). On the Tail Behavior of Record Time Distributions in a Random Record Process. *Ann. Prob.* , 7, 868–237.

Wiens, D. P. , Beaulieu, N. C. , and Pavel, L. (2006). On the exact distribution of the sum of the largest $n - k$ out of n normal random variables with differing mean values, *Statistics*, 40 (2), 165–173.

Williams, D. (1973). Renyi's Record Problem and Engel's Series. Bull. *London Math. Soc.* , 5, 235–237.

Wu, J. (2001). Characterization of Generalized Mixtures of Geometric and Exponential Distributions Based on Upper Record Values. *Statistical Papers*, 42, 123–133.

Yang, M. (1975). On the Distribution of Inter-Record Times in an Increasing Population. *J. Appl. Prob.* , 12, 148–154.

Index

Asymptotic distributions, 105, 109, 115, 116, 118, 123, 128, 186

Best linear unbiased estimates, 140
Beta function, 10, 15, 17, 66, 67, 69, 83, 206, 229
Bias, 136
Binomial Distribution, 10

Cauchy distribution, 77, 121
Characterizations, 189, 196, 214, 215
Conditional density, 50, 51, 54
Continuous distribution, 5–7, 39, 53, 75, 131, 132, 139, 146, 153, 169, 176, 177, 184, 185, 187, 190, 192, 194–196, 198, 217, 228–230
Correlation coefficients, 6, 86, 87
Covariance, 6, 76, 80, 87, 88, 91, 93, 98, 140–145, 149, 150, 154–157, 159, 164, 166, 171

Density function, 15, 18, 19, 23, 24, 29, 30, 33, 34, 41, 42, 48, 51, 54, 55, 60, 75, 77, 92, 99, 111, 115, 132, 133, 143, 164, 170, 185, 186, 196, 205, 217, 218, 220, 228, 229
Discrete distribution, 61, 63, 176
Discrete order statistics, 61, 62
Distribution function, 1, 8, 16, 17, 19, 29, 40, 51, 52, 99, 115, 117, 120, 128, 154, 164, 170, 173, 176, 178, 184–186, 189, 191, 192, 194–196, 198, 201, 203, 204, 217, 220, 228, 229

Empirical distribution, 8
Entropy, 229, 230
Estimation of parameters, 163, 214
Exponential distribution, 40–42, 44–46, 84, 89, 90, 92, 94, 113, 117, 133, 134, 156, 158,

166, 170, 171, 182, 184, 187, 194, 195, 202, 203, 205, 214, 215, 220
Extreme order statistics, 25, 115, 175, 206
Extreme value distributions, 119

Gamma distribution, 50, 85, 93, 122, 184, 216
Generalized order statistics, 217, 220

Hazard rate, 201, 205, 207, 214, 215

Independent and identically distributed, 1, 57, 141, 143–146, 148, 156, 206, 217, 228, 229

Joint probability density function, 18, 19, 30, 41, 48, 132

Laplace distribution, 24, 111, 113
Location parameter, 24, 27, 28, 164, 165
Logistic distribution, 205, 215, 216
Lyapunov theorem, 105, 106

Markov chain, 51, 55, 56, 63, 182
Median, 27, 28, 32, 35, 96, 107, 111
Midrange, 23, 28, 30, 35, 45, 169, 196
Minimum variance linear unbiased estimates, 139, 153
Moments, 75–77, 79, 80, 83–86, 88–91, 95, 96, 98, 100, 105, 141, 171, 182, 211–214, 216, 229

Normal distribution, 25, 28, 30, 95–101, 106–113, 164

Order statistics, 1, 3, 4, 6, 8, 15–19, 23, 25, 28, 29, 34, 37–46, 49–59, 61–66, 75, 76, 80, 83, 85, 86, 88–91, 95, 96, 98, 100, 105–108,

M. Ahsanullah et al., *An Introduction to Order Statistics*,
Atlantis Studies in Probability and Statistics 3,
DOI: 10.2991/978-94-91216-83-1, © Atlantis Press 2013